Pierre Gassendi and the Birth of Early Modern Philosophy

This book is the first comprehensive treatment in English of the philosophical system of the seventeenth-century philosopher Pierre Gassendi. Gassendi's importance is widely recognized and is essential for understanding early modern philosophers and scientists such as Locke, Leibniz, and Newton. Offering a systematic overview of his contributions, Antonia LoLordo situates Gassendi's views within the context of sixteenth- and early seventeenth-century natural philosophy as represented by a variety of intellectual traditions, including scholastic Aristotelianism, Renaissance neo-Platonism, and the emerging mechanical philosophy. Her work will be essential reading for historians of early modern philosophy and science.

Antonia LoLordo is Assistant Professor in the Corcoran Department of Philosophy at the University of Virginia.

Pierre Gassendi and the Birth of Early Modern Philosophy

ANTONIA LOLORDO

University of Virginia

CAMBRIDGE UNIVERSITY PRESS
Cambridge, New York, Melbourne, Madrid, Cape Town, Singapore,
São Paulo, Delhi, Dubai, Tokyo

Cambridge University Press
32 Avenue of the Americas, New York, NY 10013-2473, USA

www.cambridge.org
Information on this title: www.cambridge.org/9780521122689

First published 2007
This digitally printed version 2009

A catalog record for this publication is available from the British Library

Library of Congress Cataloging in Publication data

LoLordo, Antonia, 1972–
Pierre Gassendi and the birth of early modern philosophy /
Antonia LoLordo.
p. cm.
Includes bibliographical references and index.
ISBN 0-521-86613-8 (hardback)
1. Gassendi, Pierre, 1592–1655. I. Title.

B1887.L65 2006
194–dc22 2006005045

ISBN 978-0-521-86613-2 Hardback
ISBN 978-0-521-12268-9 Paperback

Contents

Preface

A lot of people have helped me with this book. I started working on Gassendi while writing a dissertation on his *Objections* and *Counter-Objections* to the *Meditations,* and I would like to thank my committee – Peter Klein, Tom Lennon, Pierre Pellegrin, and especially Martha Bolton – for their help and support. I wrote the bulk of this manuscript while I was a joint Fellow at Cal Tech and the Huntington Library, and I am grateful to both institutions (and the Mellon Foundation) for their generous financial support and for the lovely working atmosphere they provided. I would especially like to thank Fiona Cowie, Moti Feingold, and Roy Ritchie.

Portions of this book were presented at Dalhousie University, Simon Fraser University, the University of Florida, the University of Nebraska at Lincoln, the University of Virginia, the University of Pittsburgh, and Virginia Tech; and to the Cartesian Circle at UC Irvine, the History and Philosophy of Science Working Group at Cal Tech, the October 2001 meeting of the Midwest Seminar in the History of Philosophy, a joint meeting of the Midwest Seminar, the Centre d'Études Cartesiennes and the Centro Interdipartimentale di Studi su Descartes, and the Oxford Workshop in Early Modern Philosophy. I am grateful to all these audiences.

A number of people have provided helpful comments on the manuscript or portions thereof: Martha Bolton, Jed Buchwald, Andrew Chignell, Fiona Cowie, Jack Davidson, Raffaella De Rosa, Stewart Duncan, Moti Feingold, Dan Garber, Geoff Gorham, Alan Hajek, Chris Hitchcock, Michael Jacovides, Richard Kroll, Mike Le Boeff, Tom Lennon, Paul Lodge, Dominic Murphy, Steve Nadler, Alan Nelson, Margaret Osler,

Pierre Pellegrin, Andrew Pyle, Don Rutherford, Jorge Secada, Ed Slowik, Martin Stone, and two anonymous referees for the Press. Paul Lodge, who has offered moral support and an unreservedly sympathetic ear throughout, deserves special thanks. I would also like to thank my colleagues at the University of Virginia, who allowed me two years' leave and who have been unfailingly supportive in less tangible ways as well. Special thanks are due to Dan Devereux, Brie Gertler, Paul Humphreys, and Jorge Secada.

Finally, I would like to thank Kathryn Burns.

I am sure I have forgotten someone important. If so, it is not out of lack of gratitude, just lack of memory.

Portions of Chapters 2 and 3 appeared in the *British Journal for the History of Philosophy* (as "'Descartes's One Rule of Logic': Gassendi's Critique of Clear and Distinct Perception"). Portions of Chapters 2 and 6 appeared in *Oxford Studies in Early Modern Philosophy* (as "The Activity of Matter in Gassendi's Physics"). Portions of Chapter 10 appeared in the *Archiv für Geschichte der Philosophie* (as "Gassendi on Human Knowledge of the Mind"). I thank the editors and publishers of these journals for permission to use that material here.

References to Gassendi's Works

Animadversiones = *Animadversiones in decimum librum Diogeni Laertii, qui est de vita, moribus, placitisque Epicuri* (*Notes on the tenth book of Diogenes Laertius, which is about the life, character and opinions of Epicurus*), 1649

De apparente = *Epistolae quatuor de apparente magnitudine solis humilis et sublimis* (*Four letters on the apparent magnitude of the sun on the horizon and overhead*), 1642

De motu = *De motu impresso par motore translato* (*On the motion impressed by a moving mover*), 1642

De proportione = *De proportione qua gravia decidentia accelerantur* (*On the proportion by which heavy falling things are accelerated*), 1646

De vita et doctrina = *De vita et doctrina Epicuri* (*On the life and doctrine of Epicurus*), 1634

Disqusitio = *Disquisitio metaphysica seu dubitationes et instantiae adversus Renati Cartesii Metaphysicam et responsa* (*Metaphysical investigation, or doubts and counter-replies against the metaphysics of René Descartes, and his responses*), 1644

Examen Fluddi = *Examen Philosophiae Roberti Fluddi Medici* (*Examination of the Philosophy of the doctor Robert Fludd*), 1629

Exercitationes = *Exercitationes Paradoxicae Adversus Aristoteleos* (*Exercises in the Form of Paradoxes against the Aristotelians*), 1624

Mirrour = *The Mirrour of True Nobility and Gentility* [English translation of the 1641 *Vita Peireskii* = *Viri illustris Nicolai Claudii Fabricii de Peiresc, senatoris Aquisextiensis, vita*], 1657

Opera = *Petri Gassendi Opera Omnia in sex tomos divisa (Collected Works of Pierre Gassendi divided into six volumes)*, 1658, which includes

volumes 1 & 2: *Syntagma (= Logic, including Institutio logica; physics;* and *Ethics*
volume 3, pp. 1–94: *PES*
pp. 95–210: *Exercitationes*
pp. 211–268: *Examen Fluddi*
pp. 269–410: *Disquisitio*
pp. 420–477: *De apparente*
pp. 478–563: *De motu*
pp. 564–650 = *De proportione*
volume 4: pp. 75–480: *Observationes Caelestes*
volume 6: correspondence

PES = *Philosophiae Epicuri Syntagma (Treatise on the philosophy of Epicurus)*, 1649

Receuil = *Receuil de lettres des sieurs Morin, de la Roche, de Nevre et Gassend: en suite l'apologie du Sieur Gassend touchant la question De motu impresso a motore translato (Anthology of letters of Sieurs Morin, de la Roche, de Nevre and Gassendi: followed by Gassendi's apology concerning the question De motu impresso a motore translato)*, 1650

Syntagma = *Syntagma Philosophicum* (*Philosophical Treatise*), 1658

Vanity = *The Vanity of Judiciary Astrology, or, Divination by the stars* [English translation of a portion of the 1653 *Institutio astronomica*], 1659

Introduction

This book is both an interpretation of Gassendi's central metaphysical, epistemological, and natural philosophical views and an advertisement for their philosophical and historical interest. Historians of seventeenth-century philosophy can usually tell you that Gassendi was an atomist, an empiricist, or a mitigated skeptic, as well as an opponent of Aristotle and Descartes. They might add that he attempted to revive Epicureanism. However, few are likely to have any clear conception of the theses Gassendi articulates, the arguments he offers in their defense, or the systematic connections between them. This is an unfortunate situation, and I aim to remedy it.

There are at least two reasons why those of us who are interested in early modern philosophy and natural philosophy need to know more about Gassendi. The first is widely recognized. Gassendi's influence and the importance he was accorded by his peers and close contemporaries is unquestionable. Gassendi was a central figure in seventeenth-century philosophy and, as such, very important for the development of modern philosophical thought. He knew and was known by such figures as Descartes and Hobbes and is important for understanding Leibniz, Locke, and Newton. Were one a seventeenth-century intellectual who found Cartesianism unacceptable, Gassendi's philosophy was the obvious alternative.

Less well known, however, is the philosophical interest of Gassendi's system. Gassendi attempts to solve central problems besetting causal theories of perception; distinguishes perceptual from nonperceptual cognition in a way that idea theorists typically failed to do; argues for an explicitly antireductionist version of the mechanical philosophy; presents a radical account of the source of creaturely activity; and more. I articulate these central themes and issues in a way that makes their underlying philosophical motivations clear.

It is easy for us to think of Descartes and the Cartesian reaction to scholasticism as setting the agenda for seventeenth-century natural philosophy. I hope that, through exhibiting the intellectual situation and agenda of Descartes's chief contemporary rival, this book will have the effect of defamiliarizing the early modern philosophical landscape. We tend to think of the "new philosophers" as reacting against scholastic Aristotelianism or, in the case of later figures, Cartesianism. But even though Gassendi does write in opposition to the doctors of the schools and to Descartes, he is equally concerned with a third set of opponents – Renaissance neo-Platonists and Italian natural philosophers such as Patrizi, Telesio, and Campanella. Gassendi stands at the intersection of a number of traditions: humanism, Aristotelianism, neo-Platonism, the Italian naturalist tradition, and the new mechanist natural philosophy. Thus, coming to understand him is also coming to understand something of the great diversity of philosophical options on offer in the middle of the seventeenth century.

My concern is chiefly with natural philosophy in the broad sense. Gassendi follows the typical Hellenistic trivision of philosophy: logic, physics (otherwise known as natural philosophy or *physiologia*), and ethics. For Gassendi, logic – a discipline that has strong psychological, epistemological, and methodological components – is worth doing only insofar as it is useful, and in particular only insofar as it contributes to physics. Because Gassendi's logic is portrayed as the necessary propadeutic to physics, I include it within my treatment of natural philosophy.

The bulk of Gassendi's natural philosophy consists of detailed accounts of particular natural phenomena such as the formation of clouds and crystals. However, I concentrate on those aspects of Gassendi's natural philosophy that count as more philosophical in our sense of the term: the ontology and functions of the mind; epistemology and the theory of cognition; the metaphysics of space and the metaphysics of bodies; and the relationship between atomic and bodily explanations.

Gassendi alternately characterizes the goal of physics in terms of its contribution to ethics, as Epicurus did, and as leading us to recognize that God exists and that "the excellence and beneficence of this God should be shown reverence" (1.128b). I treat Gassendi's natural theology at some length. However, in order to have some chance of doing justice to Gassendi's physics, I omit ethics almost entirely.[1] Although Gassendi's natural philosophical work may originally have been motivated by ethical concerns, it is clear from the bulk of his natural philosophical writings and the amount of time he spent on them that natural philosophy took on a life of its own for him.

My chief focus is Gassendi's magnum opus, the *Syntagma Philosophicum*. Two considerations speak in favor of focusing on the posthumous *Syntagma* rather than the earlier *Animadversiones in decimum librum Diogenis Laertii* (*Notes on the Tenth Book of Diogenes Laertius*). First, in the *Syntagma* Gassendi writes in his own voice, while the *Animadversiones* is a commentary, albeit a rather digressive one. Although the *Syntagma* devotes a fair amount of space to reconstructing and interpreting the Epicurean view, Gassendi is careful to make clear where a revised version of Epicureanism can be embraced, and where he wishes to offer a novel view or one from another source. Second, Gassendi wanted his *Opera Omnia* to begin with the *Syntagma* and to include only the strictly philological sections of the *Animadversiones*. This decision indicates either that he was unhappy with the more philosophical aspects of the *Animadversiones* or, more likely, that he thought they had been superceded. Although Gassendi never finished the *Syntagma*, it is his most complete and systematic work.

One notable feature of the *Syntagma* is its use of a genealogical method for writing philosophy. Gassendi explicates each new philosophical question in great historical detail, summarizing and criticizing the views of major figures, before venturing any answer of his own. The use of this method is sometimes taken to indicate a conception of philosophical argument entirely different from that of contemporaries like Descartes or Hobbes – a historicist conception, on which we can neither understand nor justify philosophical positions without understanding their historical location.[2] However, the use of such a method is found in many sixteenth- and seventeenth-century texts.[3] It may well have been simply

[1] Readers interested in Gassendi's ethics should consult Sarasohn, *Gassendi's Ethics*.

[2] This is the thesis of Joy, *Gassendi the Atomist*.

[3] An arbitrarily chosen chapter of Patrizi, *Nova de Universis Philosophia*, for instance, mentions Plato, Theophrastus, Parmenides, Zoroaster, Proclus, the Chaldeans, Aristotle, Philo, Hermes Trismegistus, and Simplicius. Less than one hundred years later, the introductory chapter of Cudworth, *The True Intellectual System of the Universe*, treats the views and interpretations of Democritus, Aristotle, Plato, Leucippus, Protagoras, Posidonius, Moschus or Moses, Iamblicus, Pythagoras, Empedocles, Stobaeus,

how Gassendi thought one wrote philosophy and not the expression of any covert methodological commitments. Moreover, Gassendi's use of the genealogical method seems to me to be strikingly *anti*historicist in that it presupposes that there are grand, transhistorical questions and that everyone discussed was engaging with the same issues and with similar aims. Gassendi's discussion of human freedom, for instance, assumes that Lucretius and Suárez share a concept of freedom and have similar reasons for wanting to preserve human freedom.

What, then, *is* the significance of Gassendi's use of the genealogical method? For one thing, there is a great deal of rhetorical significance to his choice of sources to discuss or omit. It is very easy to read the *Syntagma* and think that Gassendi is attempting to provide an exhaustive historical summary. But this impression is only partly correct. Gassendi tries to recreate the whole range of classical and Hellenistic options – but he does not do the same thing for the contemporary options, leaving out most of the diversity among scholastics and treating Aristotelianism as a simple, unitary view. By doing so, he expresses a guiding assumption that the way forward will have to be found through other means than Aristotle's.

It is unfortunate that Gassendi's use of the genealogical method has made it difficult for twentieth- and twenty-first-century scholars to approach him. I hope that what follows will persuade readers to make another attempt. I begin, in Chapter 1, with an account of Gassendi's life and intellectual context, focusing on two issues that exhibit Gassendi's engagement with humanist historiography and with new natural philosophical movements: the development of his Epicurean project from biography and commentary to a positive philosophical program, and his Galileanism and his strategy for dealing with the condemnation of Galileo. Chapter 2 provides an outline of Gassendi's critiques of the opposing Aristotelian and neo-Platonist schools and of the philosophy of Descartes. It is from these critiques that Gassendi's view of matter emerges. For, he argues, examining his opponents' views shows that we can only preserve secondary causation in a theologically acceptable manner by building the active principle into matter itself.

Anaxagoras, Xenocrates, Ecphantus, Heraclides, Diodorus, Metrodorus Chius, Epicurus, Parmenides, Empedocles, Anaxagoras, Lucretius, Zeno, Chrysippus, the Chaldeans, Cicero, Seneca, Socrates, Diogenes Laertius, and Strabo. Even Sennert, *Epitome Naturalis Scientiae*, intended as a textbook, characterizes its starting question "On the Nature of Philosophy" in terms of the views of Pythagoras, Plutarch, Aristotle, and Cicero.

Chapters 3 and 4 develop an account of Gassendi's theories of perception and cognition and of the philosophical methodology put forth in the *Syntagma*'s *Logic.* In Chapter 3, I discuss his causal theory of perception. Gassendi adopts a version of the notorious Epicurean doctrine that sensation cannot lie, and this – together with the problem, common to causal theorists, of explaining how we can have different ideas of the same thing – leads to a complex and interesting theory of perceptual content. Gassendi thus offers a form of direct realism that is both more sophisticated and more explicit than the versions sometimes attributed to idea theorists like Locke. Chapter 4 explains how Gassendi's direct realism yields an account of the content of ideas that grounds the epistemology of physics. It also addresses the inferences from signs that play the dual role of providing ideas of unperceived entities and grounding probable knowledge of their existence.

Chapters 5 through 9 discuss a series of foundational issues in Gassendi's physics – space and time, the properties and motion of atoms, the structure and motion of composite bodies, the generation and life of plants and animals, and the ontology of bodies. In Chapter 5, I examine Gassendi's arguments for the existence of absolute space and time and his defense of the void against Aristotelian and Cartesian plenism. In Chapter 6, I trace the development of Gassendi's atomism in its historical context, considering how Gassendi's atoms differ from their Epicurean counterparts and the theological and physical motivations for Gassendi's revised account of the nature of atoms. On Gassendi's view, creaturely activity is built into atoms from the moment of their creation. One central problem here is how the innate activity of atoms is consistent with divine creation, conservation, and concurrence. Another is determining whether we are best off thinking of atoms as continually in motion or merely continually possessed of motive power.

My discussion of Gassendi's atomism sets the stage for an account of the relationship between the properties of compound bodies and the properties of the atoms composing them. It is not uncommon for historians of philosophy to think of mechanism as a form of reductionism about the qualities and behavior of composite bodies. This is a reasonably accurate characterization of someone like Descartes who allows no real qualities to bodies beyond the qualities of size, shape, and motion (or perhaps force) ascribed to all matter. Such a characterization, however, implies that mechanism is far less common than often thought. On such a characterization, for instance, Boyle would not count as a mechanist, nor, most probably, would Locke. For both accept the existence of emergent or super-added qualities. Nor would Gassendi count as a mechanical philosopher. For although he restricts the qualities of atoms to size, shape, and motion or the motive power underlying it, many of the properties of composite bodies cannot be reduced to such qualities. Most important among these are the various powers pertaining to generation and sensation.

Chapter 7 discusses the relationship between inanimate composite bodies and their component atoms. The relationship between Gassendi's accounts of the motion of composite bodies and the motion of atoms is awkward, although the two accounts are not, I argue, inconsistent. Chapter 8 takes on the comparatively straightforward task of documenting and understanding Gassendi's antireductionist account of life, focusing on the two central cases of generation and the sensitive powers of the corporeal soul. The crucial issue for antihylemorphic theories of generation is to explain how the complex structure of a mature organism can develop out of undifferentiated matter, and Gassendi, like later preformationists, ends up ascribing a great deal of preexisting structure to seeds. In Chapter 9, I analyze the way Gassendi reinterprets the traditional ontological categories of substance, nature, and accident in corpuscular terms. So doing allows us to elicit a general ontology on the basis of the more localized accounts of the previous chapters.

I end, in Chapter 10, with an account of the status and content of our knowledge of God and the incorporeal soul. My account revolves around two issues: first, how Gassendi accommodates cognition of the incorporeal within his radically empiricist theory of cognition and, second, how he deals with the ontology of the incorporeal given his corpuscularian understanding of the categories of substance, nature, and accident. Once we have a sophisticated understanding of Gassendi's ontology and epistemology of the mind, we are in a position to address the vexing issue of the relationship between faith and reason, which has been a central topic in recent work on Gassendi.[4]

[4] As well as Sarasohn, there are three other major recent books on Gassendi, and all of them treat the dialectic of faith and reason at some length: Bloch, *La philosophie de Gassendi*; Brundell, *Pierre Gassendi*; and Osler, *Divine Will and the Mechanical Philosophy*.

1

Gassendi's Life and Times

Pierre Gassendi was born – as Pierre Gassend, son of the peasant farmer Anthoine Gassend and his wife Françoise Fabry – on January 22, 1592, in the village of Champtercier, near Digne in Provence. This was the year of Montaigne's death. He attended the Collège de Digne from 1599 to 1607 – where he learned, primarily, Latin[1] – and the Faculté d'Aix beginning in 1609, studying philosophy with Père Philibert Fesaye. He also followed a course of theology that included Greek and Hebrew.[2] Fesaye was a Carmelite, and the *ratio studiorum* of the Carmelites refers to Aquinas, Toletus, Averroes, and the Carmelite doctor John Bacon or Baconthorp. Baconthorp attacked intelligible species; denied the univocity of being; held that universals precede the action of the intellect and that external objects are intelligible per se although understanding them requires an agent intellect; equated essence with quiddity; and maintained a formal distinction between essence and existence.[3] Gassendi adopted none of these doctrines, save the rejection of intelligible species.

Gassendi was recognized as an exceptional student from early on, and received his doctorate in theology in 1614, at the age of 24, at which time he took the four minor orders of the church and became the theological canon of Digne Cathedral.[4] He kept this job until he was promoted to provost in 1634, after some legal wrangling. In 1616, Gassendi was ordained to the priesthood. In 1617, he was offered the chairs in

[1] La Poterie, "Memoires," 215.
[2] Bougerel, *Vie de Pierre Gassendi*, 7.
[3] Armogathe, "L'Enseignement de Pierre Gassendi."
[4] La Poterie, "Memoires," 216–17.

both philosophy and theology at the University of Aix; he took philosophy, leaving the theology chair to his old teacher Fesaye. At this time, Gassendi was living at the house of the astronomer Joseph Gaultier, "who," Gassendi wrote, "had no difficulty in equaling all the ancient and modern philosophers and mathematicians."[5] Gaultier also provided lodging for Jean-Baptiste Morin, with whom Gassendi would later have a protracted quarrel, and their fellow astronomer Ismail Bouillau, who would become an important correspondent of Gassendi.[6] It was from Gaultier that Gassendi learned much of his astronomy. The two observed a comet together in 1618, and eclipses of the moon and sun, respectively, in 1620 and 1621.[7]

It was in 1617 also that Gassendi met his future friend and patron, Peiresc. Nicolas-Claude Fabri de Peiresc was an influential humanist and antiquarian, known across Europe for his erudition and from his voluminous correspondence with all sorts of intellectuals. Peiresc had interests in numismatics, botany, astronomy, antiquities more generally, and books of all kinds. Indeed, Peiresc is spoken of as an ideal of the late humanist type.[8] He was also, Gassendi tells us,

> studious of Mechanics, or Handi-Crafts; for which cause, there was never any famous Workman that went that way, but he entertained him at his House, and learnt of him many works of mysteries of his Craft; for he would keep him with Diet, wages, and gifts, and make much of him for months and years together. (*Mirrour* 186)

Gassendi goes on to tell us that Peiresc was an admirer of Bacon and an opponent of scholastic doctrines of nature, both respects in which he and Gassendi were of similar mind.

Peiresc was an important figure in Gassendi's early astronomical career. Peiresc had spent the winter of 1599–1600 in Padua, where he attended lectures by Galileo, and after hearing about Galileo's telescopic discovery of the Medicean stars (the moons of Jupiter) in 1610, he had an observatory built and hired Joseph Gaultier to work there. One of Gaultier's first projects was to compute the times of the revolutions of the four moons, and in order to help Gaultier do the computations more

[5] Letter to de Pibrac of April 8, 1621. Quoted by Bougerel, *Vie de Pierre Gassendi*, 13.

[6] Bougerel, *Vie de Pierre Gassendi*, 9.

[7] At least, Bougerel, *Vie de Pierre Gassendi*, 10–11, says that the observations were done with Gaultier, although the record in Gassendi's *Observationes Caelestes* – the posthumous compendium of his astronomical observations – makes no mention of this (4.77a).

[8] For this claim and an excellent account of Peiresc's life, reputation, correspondence networks, and significance, see Miller, *Peiresc's Europe*, passim.

quickly Peiresc hired Morin and Gassendi as helpers (*Mirrour* 143f). Peiresc was to become Gassendi's patron and close friend. It was from him that Gassendi acquired his first telescope, one that had been given to Peiresc by Galileo. It was also through Peiresc's introductions that Gassendi became acquainted with the circle of thinkers around Marin Mersenne – a circle that largely shared Gassendi's admiration of Galileo and that was to be extremely influential for Gassendi's later career.

Gassendi stayed at Aix until 1622 (or perhaps 1623[9]), when the university and its curriculum were placed under Jesuit control and Gassendi had to leave.[10] All the non-Jesuit faculty were dismissed; Gassendi's departure had nothing to do with any particular philosophical beliefs nor, indeed, had he yet published any of those beliefs. After he left Aix, Gassendi returned to Digne, and the next year Book I of the projected seven books of his *Exercitationes Paradoxicae Adversus Aristoteleos (Exercises in the Form of Paradoxes Against the Aristotelians)* was published. Gassendi explained his motivations for writing the *Exercitationes* as they developed during his time teaching at Aix:

> I always made sure that my students could defend Aristotle properly. But at the same time, I also provided as appendices doctrines that would undercut Aristotle's dogmas. Indeed, given the place, the characters and the times, it was necessary to do the former. But not to omit the latter was a matter of candor because those doctrines provided true reason for withholding assent. (3.100)

Exercitationes I was a collection of these more critical parts of his lectures. As we can see from the language of withholding assent, Gassendi's own philosophical allegiances at the time were to skepticism, if anything. He wrote that while he was becoming disillusioned with Aristotelianism, he "began to examine the doctrines of other sects to find out whether they perhaps might offer something sounder. Although I found perplexities everywhere, none of the doctrines impressed me more than the lauded *akatalepsia* of the Academics and Pyrrhonians" (3.99). Although Gassendi's skeptical sympathies diminished significantly over time, skepticism remained an important influence on him.

In October of 1623, Gassendi traveled to Paris, where he met Marin Mersenne and, through him, a number of other prominent intellectuals.[11] Gassendi and Mersenne became good friends, and it is

9 For difficulties determining the date, see Joy, *Gassendi the Atomist*, 25 n. 2.

10 Brundell, *Pierre Gassendi*, 1, argues that this was the result of a rather late implementation of the Council of Trent's call for a reformation of seminaries.

11 La Poterie, "Memoires," 236.

said that until Mersenne's death Gassendi celebrated Mass with him at his convent whenever he was in Paris. Through Mersenne's intermediacy, Gassendi met (at one point or another) the mathematician Gilles Personne de Roberval, the poet Jean Chapelain, Hobbes and the Cavendish family, Grotius, and perhaps Pascal. Gassendi apparently became friends with Hobbes during his time in Paris in the 1640s, although the two had met previously. Hobbes wrote in his autobiography that he later "returned again to France where he could study knowledge more securely with Mersenne, Gassendi and other men," and Gassendi, along with Mersenne, wrote a commendatory letter printed with the third edition of *De cive.*[12] Lisa Sarasohn has argued that there is significant influence between the two men's political theories as well as their natural philosophies.[13] Samuel Sorbière, a disciple of both Gassendi and Hobbes in turn, tells us that when Gassendi was given a copy of *De corpore* on his deathbed, he greeted it with a kiss.[14] Although Sorbière is by no means a trustworthy source, it is also worth noting his report of Hobbes's claim, concerning the *Fifth Objections* and *Counter-Objections,* that Gassendi "never appeared greater than when beating back the ghosts" of metaphysical speculation.[15]

Soon after his journey to Paris, Gassendi – with the encouragement of his Genevan friend Eli Diodati – first wrote to Galileo, telling him rather effusively, even by the standards of the day, that he had long known and admired his work and was in full agreement with him concerning Copernicanism. Around this time, Gassendi abandoned the *Exercitationes,* although he kept the finished but unpublished manuscript of Book II, "On the dialectic of the Aristotelians." Lynn Joy notes two possible explanations for abandoning the *Exercitationes.*[16] One line of explanation emphasizes the significance of Gassendi's Paris trip of 1624–5 and the conversations with Mersenne and others that might well have led him to realize that Book II would greatly offend some powerful people. (Some have suggested that the recent condemnations of Jean Bitaud, Antoine

[12] Hobbes, *Opera philosophica quae latine scripsit,* 1.xiv.

[13] Sarasohn, "Motion and Morality," argues that Hobbes's psychology was influenced by the views Gassendi was developing in the 1630s, and that Gassendi's turn away from thoroughgoing materialism in the early 1640s was spurred at least in part by reaction against Hobbes.

[14] Sorbière's unpaginated preface to the *Opera,* twenty-second page.

[15] Ibid., eighteenth page.

[16] Joy, *Gassendi the Atomist,* 32–7.

de Villon, and Étienne de Clave had some force in this too, although the main basis of their condemnation was that their chemical atomism was "false, audacious and contrary to the faith."[17]) The second is that Gassendi simply came to realize that Patrizi had already made basically the same case in his anti-Aristotelian work, so that Book II would be redundant.[18] Joy suggests the interesting further consideration that by this time, although Gassendi was still committed to opposing Aristotelianism, he was no longer enamoured of the skeptical mode of arguing he had earlier used. One possible reason for this disenchantment, Joy argues, is the influence of the antiskeptical doctrines of Mersenne's *La vérité des sciences*, which made clear to Gassendi that one could oppose Aristotelianism without giving up the possibility of knowledge altogether. Whatever the real reason, none of Gassendi's later works evince the thoroughgoing skepticism found in the *Exercitationes.*

In 1626, Gassendi returned to Digne. His correspondence with Peiresc shows that, by this time, he had already begun working on Epicurus.[19] He stayed in Digne for a few years but in 1628 returned to Paris, where he met the *libertin érudit* François Luillier and the circle of writers surrounding the Dupuy brothers. The Dupuys had been corresponding with Peiresc for years, so this was probably through Peiresc's intermediacy. Since the death of their uncle, the historian Jacques Du Thou, Jacques and Pierre Dupuy had administered and added to his library, and it was in this library that Gassendi would carry out much of his research on Epicurus.

In December, Gassendi and his friend François Luillier traveled to Holland for several months. In July, he met with Isaac Beeckman and, like Descartes, was greatly impressed by him: Writing to Peiresc, he called Beeckman "the best philosopher I have yet met."[20] Beeckman knew Gassendi as the author of the *Exercitationes*, a doctor of theology, and a canon of Digne. The two discussed music and motion, and Beeckman wrote in his journal that Gassendi "approved" what he heard and "seemed

[17] Villon and De Clave replaced the four Aristotelian elements with the five chemical elements – salt, sulfur, mercury, earth, and water – and mounted an attack on Aristotle in their fourteen theses. Kahn, "Entre atomisme, alchimie et théologie." For the effect of the condemnation, see also Spink, *French Free-thought from Gassendi to Voltaire*, 89–90.

[18] Rochot, "Vie et Caractère," argues that this is implausible because Gassendi would almost certainly have known about Patrizi's work *before* 1624 and *Exercitationes* I borrows numerous arguments and rhetorical tropes from Patrizi.

[19] Tamizey de Larroque, *Lettres de Peiresc*, 4.178–81.

[20] Ibid., 4.198–202.

to hear with joy and admiration."[21] A few, not terribly interesting, letters followed.

Gassendi's next published works were the *Phaenomenon rerum*, printed (poorly, he thought) in Holland in 1629 and reprinted in France the next year under the title *Parhelia sive soles quatuor spurii* (*Parhelia or four spurious suns*); and the *Examen Philosophiae Roberti Fluddi Medici* (*Examination of the Philosophy of the doctor Robert Fludd*), written at Mersenne's behest to combat the philosophy of the English doctor Robert Fludd. Gassendi's second astronomical publication, *Mercurius in sole visus* (*Mercury seen on the face of the sun*), the record of his confirmation of Kepler's 1629 prediction of the transit of Mercury across the face of the sun, was published in 1632. The next year or two were spent in Provence, where the Epicurus project appears to have continued; at least, Gassendi made repeated visits to Peiresc and his library. The relative serenity of this period was disturbed only by news of the condemnation of Galileo.

In 1634, the second of Gassendi's occasional works on other philosophers was written and published, this time at his patron Peiresc's request: *Ad librum D. Edoardi Herberti de Veritate Epistola* (*A Letter concerning Edward Herbert's book On Truth*), which addressed Herbert's antiskeptical arguments. Among other things, Gassendi objects to Herbert's fourfold distinction between kinds of truth; his reliance on innate "common notions"; his claims about what is universally agreed upon; and his careless treatment of sense perception. Many of the important themes and interests of Gassendi's later work appear in his criticisms of Herbert of Cherbury. Particularly notable are Gassendi's attacks on innate ideas and his concern with the details of sense perception.

The next year, Gassendi wrote the first of four letters that would be published together as *De apparente magnitudine solis humilis et sublimis* (*On the apparent magnitude of the sun on the horizon and overhead*). The first letter was written to his friend Gabriel Naudé from Aix on December 5. A second letter, written to Fortunio Liceti from Aix on August 13, followed in 1640; the third and fourth appeared in 1641. The four letters *De apparente* deal with problems parallel to the moon illusion, and Gassendi offers an atomist explanation of them.[22]

Peiresc had supported Gassendi throughout this period. The two were close, and Peiresc's death on June 11, 1637, seems to have come as rather a

[21] Beeckman, *Journal tenu par Isaac Beeckman de 1604–1634*, 3.123.
[22] For more on *De apparente*, see Joy, *Gassendi the Atomist*, 106–29.

shock. (Insult was added to injury when Peiresc's nephew refused to give Gassendi the books and mathematical instruments bequeathed to him in Peiresc's will.[23]) In memory of Peiresc, Gassendi wrote a biography describing not only their friendship but also the work in natural philosophy they had done together. This is thought to be the first instance of a new genre – the biography of a scholar.[24] The *Vita Peireskii* was widely read: The first publication in 1641 was followed by Dutch printings in 1650 and 1655 and, in 1657, an English translation printed under the name *The Life of Peireskius* or *The Mirrour of True Nobility and Gentility.* Apparently, biography was to Gassendi's liking: he would later write a full-length biography of Tycho Brahe and much shorter biographies of the astronomers Copernicus, Peurbach, and Regiomontanus.

The loss of Peiresc was both the loss of a friend and the loss of a patron. Louis de Valois, then governor of Provence, soon replaced Peiresc as Gassendi's patron. They never achieved the same degree of intimacy, however, and de Valois was not a member of the republic of letters in his own right, as Peiresc had been. However, as governor he was a rather more powerful figure than Peiresc: De Valois's patronage reflected Gassendi's increased philosophical status.

In 1640, Gassendi wrote the first two letters *De motu impresso par motore translato* (*On the motion impressed by a moving mover*), addressed to Pierre Dupuy. They were published in 1642 but circulated widely beforehand. In these letters, Gassendi puts forth a version of Galilean relativity to defend Copernicanism from the objection that heavy bodies would not fall as they do if the earth were moving. The cautious Gassendi expressed this argument not as a defense of Copernicanism but merely as showing that one common argument against Copernicanism was wrong.

Despite Gassendi's care on this last point, the publication of *De motu* occasioned some controversy. Both Morin and the Jesuit Pierre Cazré published attacks on *De motu* and its allegedly Copernican implications. Gassendi replied to Cazré in the first letter *De proportione qua gravia decidentia accelerantur* (*On the proportion by which heavy falling things are accelerated*)[25] on December 8 that year. He also replied to Morin, in 1643's third letter *De motu.* This was not, however, published until 1649, in the

[23] La Poterie, "Memoires," 224, gives a very brief account of the disapproval this was generally met with, and of Peiresc's nephew's insinuations.

[24] Miller, *Peiresc's Europe*, 16–48.

[25] Two more letters were written, and the three were published together in 1646.

Apologia in Morini Librum, and even then against Gassendi's wishes. While the *De motu* controversy was going on, Gassendi was also writing what is now his most widely read work, the *Fifth Objections* to Descartes's *Meditations*, which would be followed by the much longer *Counter-Objections*. I discuss both the arguments and the circumstances of composition of these counter-Cartesian works in Chapter 2.

Alphonse de Richelieu, cardinal-archbishop of Lyon and brother of the already-deceased Prime Minister Cardinal Richelieu – who had known Gassendi since the 1620s when he was archbishop of Aix – named Gassendi professor of mathematics at the Collège Royal in 1645.[26] The chair of mathematics was given him in recognition of his prowess in astronomy, one of the branches of mixed mathematics, rather than any concern or talent for pure mathematics. Gassendi officially kept this position until his death, although on his physicians' advice he stopped teaching earlier.[27] In 1647, the first public installment of the Epicurus project was published at Lyon under the title *De vita et moribus Epicuri* (*On the life and character of Epicurus*). This was intended both to demonstrate the impeccable personal morals of Epicurus and to render his doctrines consistent with the Catholic faith. Gassendi's *Institutio astronomica*, together with his inaugural lectures at the College Royal, was also published in that year. The *Institutio astronomica* was widely reprinted, at, for instance, London in 1653 and again in 1683, the Hague in 1656, and Amsterdam in 1680.

Part of the *Institutio astronomica* was published in 1658 in English translation as *The Vanity of Judiciary Astrology*. There Gassendi is concerned to argue against two different sorts of astrology: the kind that predicts the weather on the basis of the stars and the kind that predicts the fortunes of particular men from their genitures or nativities. Neither has any real predictive power, according to Gassendi. The motions of the stars are not – contrary to vulgar opinion – the causes of tempests and mutations in the air but only signs.[28] Predictions of men's fortunes on the

[26] Sarasohn, *Gassendi's Ethics*, 12.

[27] Bougerel, *Vie de Pierre Gassendi*, 394.

[28] Gassendi gives a number of arguments for this claim. For example, the fixed stars rise a month later in the year than they did in the times of the ancient Greek philosophers, but the seasons remain unchanged and aligned to the sun. For another example, Sirius acts as a sign of heat to us but in the antipodes it is a sign of cold, so the appearance of Sirius cannot itself be the cause of either heat or cold – and so on for all the fixed stars. In any case, Gassendi notes, even in France the connection between the stars and the weather is not exact, for weather is far more variable than the predictable heavens.

basis of their genitures fare even worse. Gassendi notes that people have an unfounded faith in judiciary astrology because of their tendency to remember successful predictions and forget failed ones. Moreover, there are pragmatic objections to the use of nativity schemes because they are not conducive to either peace of mind or sensible behavior. If an early death is predicted, a man may become nervous and make himself sick; if long life is predicted, he may become reckless. Indeed, Gassendi writes, "we have many stories both Ancient and recent that testify, that those men have generally been most unhappy, who confided in the Astrologers promising them very great happiness" (*Vanity* 14.87).[29]

Gassendi also mounted a more theoretical attack on judiciary astrology, laying down a general rule for the knowledge of future events: "[W]hatsoever, doth import the knowledge of any effect to come, ought to be either the necessary Cause of that particular effect, or which being posited, such an effect doth alwaies follow; or as a necessary Signe, or which being given, such an Event doth always succeed" (*Vanity* 14.2). The first clause is satisfied by things like the conjunction between the sun's approach to the vernal equinox and the budding of plants; the second, by the morning twilight and the rise of the sun. The connections alleged between nativities and the fortunes of men fit neither of these two cases. Nor, Gassendi goes on to argue, do they satisfy even the weaker standards of reasonable conjecture where "many Observations concurr to attest, that such a Cause, or Signe is more frequently attended on by that particular effect, than not, or that that effect doth more usually succeed upon that Cause, and after that Signe, than upon any other" (*Vanity* 4.2). This is an application of the theory of signs, which is crucial to Gassendi's epistemology.

Gassendi was in Paris in 1648 and attended Mersenne on his deathbed.[30] Shortly thereafter, he returned to Provence, in part for the sake of his health: He had suffered from pulmonary and respiratory problems since his youth. The second, rather larger, installment of the Epicurus project came out in 1649 in three volumes as *Animadversiones in decimum librum Diogenis Laertii (Notes on the tenth book of Diogenes Laertius,* which is the chapter about Epicurus in *Lives of Eminent Philosophers).* The original publication of the *Animadversiones,* which sold out quickly in

[29] That is, page 87 of Chapter 14. Page numbers in the edition I consulted are nonstandard; for instance, Chapter 4 starts over at page 1 but numbers for Chapters 1–3 run consecutively.

[30] Rochot, "Vie et Caractère," 45.

Paris, contains as an appendix the *Philosophiae Epicuri Syntagma*.[31] Samuel Sorbière apparently translated the *PES* into French, but the translation was never published and has since been lost.[32]

Gassendi's health improved considerably while he was living in Digne and in Toulon (where Louis de Valois had moved as a result of the recent troubles in Aix), despite Morin's prediction of impending death.[33] By February 5, 1650, he was well enough to ascend a nearby mountain to repeat the experiment of the Puy-de-Dôme with Bernier.[34] For the next three years, Gassendi continued to live and work relatively quietly in Provence. On September 25, 1652, Queen Christina of Sweden wrote to Gassendi, saying that she consulted him "like an oracle of truth for clarifying her doubts," wondering if he would mind taking the trouble to instruct her ignorance, and so forth.[35] A letter from her chief physician followed, broaching the subject of Gassendi coming to Sweden more explicitly. Gassendi excused himself on the grounds of advanced age (he was then sixty), continual infirmities, and the fact that he was accustomed to living in a more temperate climate than that of Sweden. (He did not, tactfully enough, mention Descartes's unfortunate demise.) Indeed, he had been advised that even the harshness of the Parisian climate should be avoided.

Nevertheless, in May of 1653, Gassendi returned with Bernier to Paris and took up residence at the house of his new patron, Henri-Louis Habert de Montmor. This was a sad year: Gabriel Naudé had died in July, and Louis de Valois, in November. But in 1654, along with the biography of Tycho Brahe, Gassendi produced a number of minor works: *De sestertiorum moneta nostra expressorum Abacus*; *Romanum Calendarium compendiose expositum,* which treats the calendar up to Gregory XIII's reformation; *Notitae*

31 For the claim that the *Animadversiones* was quickly sold out, see a letter of May 16, 1649, from John Pell to Charles Cavendish, quoted in Hervey, "Hobbes and Descartes," 78.

32 Rochot, "Vie et Caractère," 89.

33 During their quarrel over Copernicanism, Morin cast a horoscope of Gassendi showing that although Gassendi had "a good mind, appropriate for all the sciences he applied himself to," the malign influence of Saturn also made him "inclined to dissimulate" and "easily irritated" – characteristics that, he says, Descartes, "a man of knowledge and great reputation," had also encountered in his dealings with Gassendi. He also predicts that Gassendi would die from lung troubles after two years' illness. It took longer than that, but Morin's prediction of sickness and lung problems were accurate, although, of course, Gassendi had a history of such illness well before Morin cast his horoscope (*Receuil* 9).

34 For a description of this experiment, see Chapter 5.

35 Bougerel, *Vie de Pierre Gassendi*, 368.

Ecclesiae Diniensis; *Manuductio ad Theoriam seu partem speculativam Musicae*; and one of his two, anonymous, French publications, *Sentiments sur l'Eclipse qui doit arrivée le 12 du mois d'août prochain, pour servir de réfutation aux faussetez qui ont esté publiées sous le nom du docteur Andreas*.[36] An alarmist pamphlet, predicting that dire consequences would attend the upcoming eclipse, had become well known in Paris. Gassendi's sixteen-page response, written in an attempt to calm the people, explains to his readers that an eclipse is just a very short night, no more likely to cause harm than any normal night, and that only God knows when the end of the world will come. During this period, Gassendi was in the habit of attending a Saturday academy of mathematicians, along with such old friends or acquaintances as Bouillau, Pascal, Roberval, and Girard Desargues, and it may have been this group who inspired him to write *Sentiments*.[37]

Gassendi fell seriously ill that winter, and – although his health recovered in the spring of 1655 – he grew sick again that August. Guy Patin and several other physicians attended him. Bougerel tells us that there was some dispute about when they should stop bleeding him, but in the end, he was bled fourteen times.[38] Gassendi died on October 24, at Montmor's Paris house and was buried three days later at Saint-Nicolas-des-Champs in the Montmor family chapel. His last words, at least according to Sorbière, were "vides quid sit hominis vita" (behold what is the life of man).[39]

He left his personal effects and a reasonable sum of money to his secretary, Antoine de La Poterie, in recompense for unpaid wages. He left the choice of a few of his books to Montmor and to his "good and dear friend" Jean Chapelain, and the rest to his secretary. The books inventoried after Gassendi's death included Proclus on Plato, in Greek and Latin; Digby's *Demonstratio immortalitatis animae rationalis*; a Bible in Hebrew and Latin; the *Méthode pour convertir ceux qui sont hors l'églize* of Armand-Jean de Richelieu; Thomas's commentary on Boethius' *Consolation of Philosophy*;

36 The other is the *Discours sceptique sur la passage du chyle & sur le mouvement du coeur*. This anonymously published work is somewhat sympathetic to Harvey's account but ultimately rejects it. Gassendi there claims to have observed the septum between the two chambers of the heart. (In the *Examen Fluddi*, Gassendi had earlier attacked Fludd's use of the circulation of the blood to defend a circulation within the macrocosm.) French, *William Harvey's Natural Philosophy*, 328–34.

37 Bougerel, *Vie de Pierre Gassendi*, 408.

38 Ibid., 409–10.

39 Ibid., 412.

Fineus' *In Euclideum*; Plutarch's *Lives*; the Greek text of Plotinus's *Enneads*; and Ficino's edition and translation of Plato's works.

Gassendi also left behind a large number of papers. In a small locked armoire were found a folio volume called "Epistole," one called "De rebus celestibus," and another titled "Liber duo dessimus de Universo"; two small treatises called "De musica" and "Adversus Aristoteleos" (presumably, the second part of the *Exercitationes Paradoxicae Adversus Aristoteleos*); and a bag of money. Gassendi's remaining manuscripts and other possessions were brought from Montmor's country house: These included a Greek Aristotle, a copy of the *Animadversiones*, and a number of manuscripts – notably some labeled "Syntagma philosophicum" and "Syntagmatis philosophici pars secunda" – as well as a red leather telescope given to him by Galileo and some other astronomical instruments. Montmor took the red morocco telescope, while La Poterie received the remainder of Gassendi's astronomical instruments. All the manuscripts were left to Montmor, with the suggestion that Gassendi's disciple Bernier and his secretary La Poterie should help in organizing them. His few household goods were left to his sister Catherine.[40]

After Gassendi's death, Sorbière and Montmor began constructing a finished text of the *Syntagma* from these manuscripts so that they could publish an *Opera Omnia*. The completed *Opera* came out in six folio volumes in 1658. A number of English publications of Gassendi came out in the next few years as well: the very popular *Vanity of Judiciary Astrology* in 1659; the *Institutio logica* and *Philosophiae Epicuri Syntagma* in 1659, and – after the appearance in 1678 of Bernier's only slightly shorter *Abrégé* of the *Syntagma* – a translation of Bernier's version of the *Ethics*, under the name *Three Discourses of Happiness, Virtue and Liberty*. Two other widely read English works contain partial translations and partial, not always accurate expositions of Gassendi's views: Walter Charleton's 1653 *Physiologia Epicuro-Gassendo-Charletoniana* and Thomas Stanley's three-volume *History of Philosophy* of 1655–6. Both contain quite a few non-Gassendian elements as well. Gassendi's *Opera Omnia* was reprinted in 1727 at Florence; this is the same text as the 1658 *Opera*, with a few corrections and different page numbers. Although the *Opera* was reprinted in facsimile in 1964, there is, unfortunately, no contemporary critical edition of his works.

[40] Gassendi's will and the inventory of his belongings, together with useful annotations, are published as Fleury and Bailhache, "Documents inédits sur Gassendi."

Bougerel and La Poterie, Gassendi's first biographers, more or less agree about his character. They describe him as gentle, easy-going, sociable, candid, and – importantly for someone dependent on patronage and the good will of other intellectuals – humble. In the same vein, one of Hobbes's letters to Gassendi ends with Hobbes expressing his desire "to imitate, so far as I can, your virtuous way of life, and to distinguish [it] from the false, pompous, and histrionic life . . . of those who, in spite of everyone else, desire to be the only authorities."[41] Bougerel and La Poterie also argue that Gassendi's personal piety was unexceptionable, noting, for instance, that he held Mass every Sunday of his life except when he was too sick to go to church. (This was, mind you, quite a lot of the time.) One might suspect the accuracy of these biographers; certainly some of the stories they recount – the sermons Gassendi preached to his friends as a small child, for instance – are clearly myth making. However, a similar view of Gassendi is evidenced in less partisan texts such as Gabriel Daniel's *A voyage to the world of Cartesius*:

> We landed in *Gassendi,* a Seat extraordinary fine and very apposite, and such in a Word as an *Abbot,* like *Monsieur Gassendus,* could make it, who wanted for neither Genius, Art nor Science, and who had no use for his Revenues, in gaming treating and living high. The Lord of the Mannor was then absent, whom we should have been glad to have waited on, since we heard that he still continu'd his Civility and Moderation, which were his Natural Endowments. And though formerly there were some Misunderstandings betwixt him and *Cartesius,* yet he always very obligingly, and with a Mark of Distinction, entertain'd the *Cartesians* that came to pay a Visit, and especially *Father Mersennus,* who was his peculiar Friend. He was a Man that equall'd *M. Descartes* in capacity of *Genius,* excell'd him in the reach and extent of Science, but was less heady and conceited. He seem'd somewhat a *Pyrrhonist* in Natural Philosophy, which in my Opinion is becoming enough of a Philosopher, who provided he looks into himself, must know by his own Experience the Limits of a Human Understanding, and the short Sightedness of its Views.[42]

Indeed, it is notable that Gassendi's life was largely free of quarrels and disputes, and despite the fact that he held views that could easily have been seen as heterodox Gassendi never encountered any difficulties with church authorities.[43] This shows how carefully Gassendi formulated his views and how important the patronage networks he belonged to were.

[41] Hobbes, *The Correspondence,* 185.

[42] Daniel, *A voyage to the world of Cartesius,* 99.

[43] There are a few exceptions to this. Gassendi's argument with Morin over Copernicanism was fairly fraught, and Gassendi was involved in some wrangling over Peiresc's will and over his promotion to provost of Digne.

The Epicurus Project

Gassendi's work on Epicureanism resulted in a number of publications within his lifetime – *De vita et moribus Epicuri libri octo* (1647), *Animadversiones* (1649), and the *Philosophiae Epicuri Syntagma* (1649) – but its most developed form, the *Syntagma Philosophicum*, was never finished. Rather, it was put together after Gassendi's death by Sorbière and Montmor. A large portion of the manuscript they used is now missing, but much survives, distributed among the Bibliothèque Nationale (Paris), the Bibliothèque Municipale (Tours), and the Bibliothèque Laurentienne (Florence).[44] The fact that Gassendi's major work was never finished or published during his lifetime presents certain scholarly complications. In the case of apparent inconsistencies, we often have no way to determine whether these are *merely* apparent, whether they indicate a change in views over time, or whether they actually reflect deep tensions within his thought. I have tried to adjudicate between these possibilities on a case-by-case basis. A critical edition of Gassendi's writings – one that gave clear dating of the various sections of the *Syntagma*, indicated changes made in manuscript, and printed a larger selection of letters than those Sorbière and Montmor included in the *Opera* – would be immensely useful. Despite the absence of such an edition, we can trace some of the development of Gassendi's Epicurus project by way of the dates of composition and revision that Olivier Bloch provides.

The first mention I know of the Epicurus project is in a letter to Peiresc of April 25, 1626, although this amounts to little more than a demonstration of interest.[45] In 1628, Gassendi remarked that he had written an apology for Epicurus to be added as an appendix to some future volume of the *Exercitationes* (6.11b), and the next year Gassendi told Beeckman that he would send him "his Epicurus" soon (6.26b). By 1631, the project had expanded. On April 18, Gassendi wrote to Peiresc that his "Apology for Epicurus" would be followed by an account of Epicurus' doctrine, divided into canonic, physics, and ethics (although he omits a description of the ethics). At this point, Gassendi is obviously aware that he will need to do some work in order to make Epicureanism acceptable to his

44 For a fuller account of surviving manuscripts, see Bloch, *La Philosophie de Gassendi*, 498–501. More recent information can be found in Centre International d'Études Gassendiennes, *Catalogue: Pierre Gassendi*. A recently discovered manuscript is described by Palmerino, "Pierre Gassendi's *De Philosophia Epicuri Universi* Rediscovered." The manuscript is British Library MS Harley 1677.

45 Tamizey de Larroque, *Lettres de Peiresc*, 178–81.

readers. After a seven chapter account of Epicurus' life and moral standing, Gassendi planned a five chapter book on canonic, giving an account of the various criteria for judging truth and on the canons of words, sense, anticipation, and physics. Physics was to be divided into four books – *De natura, De mundo, De sublimis,* and *De humilibus.*[46] After compiling extensive notes and then deciding to start anew, reordering them, Gassendi worked on this project from 1636 until the early 1640s. The resulting work is known as *De vita et doctrina Epicuri* (*On the Life and Doctrine of Epicurus*). Various copies of this manuscript were made, and the full text is present in different hands in various libraries.[47]

In the end, the *De vita et doctrina* project was used as the basis of two published books. (Some material written for *De vita et doctrina* also survives in the *Syntagma* without ever having been published in Gassendi's lifetime.) The first was *De vita et moribus Epicuri libri octo* – the seven chapters of the 1631 outline had grown to eight. This was published in 1647, and followed two years later by the second half of the original project, the *Animadversiones in decimum librum Diogenis Laertii.* In the *Animadversiones* Gassendi provides a Greek text of Book X of Diogenes – including three long letters of Epicurus on physics, ethics, and astronomy – and his own Latin translations, together with more than sixteen hundred pages of philosophical and philological commentary. Gassendi's text is based on a number of sources, including Henri Estienne's edition as revised by Casaubon, and transcriptions made by the Dupuy brothers. He worried about errors and discrepancies among the various versions of the text available to him; establishing the text from what was available to him in Du Thou's library appears to have cost him a great deal of time and effort (*Animadversiones* 1.102–3).[48] The Greek text and Latin translation are reprinted in volume 5 of the *Opera* under the title *Animadversiones,* but with the commentary reduced to a little over a hundred pages of strictly philological material.

Gassendi seems to have never been entirely happy with the *Animadversiones.* Because it is intended as a commentary, his own arguments are often difficult to distinguish from exegesis. The mixture of philological and philosophical commentary can be difficult and tedious for

46 Ibid., 249–52.

47 Bloch, *La Philosophie de Gassendi,* and Brundell, *Pierre Gassendi,* describe the manuscript remains of *De vita et doctrina* carefully; however, the one part they describe as missing – the preface, *De philosophia Epicuri universe* – is British Library MS Harley 1677.

48 See Joy, *Gassendi the Atomist,* 74–7, for an account of the sources, quality, and reception of Gassendi's Greek text.

contemporary readers, who are more likely to be primarily interested in Gassendi's philosophical views. It is somewhat hard to discern where Gassendi is agreeing with Epicurus and where he is merely reporting his views. Only in cases where Epicurean atomism clearly needs major amendment to be acceptable to a Christian reader does Gassendi explicitly dissent. And even though Gassendi never makes the sort of grand claims about the greater theological acceptability of Epicureanism to Aristotelianism that Lipsius made about Stoicism and Ficino about Platonism, there is clear historical precedent for making fairly major amendments of an ancient philosophy in the course of reviving it.

It seems that Gassendi never thought of the *Animadversiones* as entirely finished. Indeed – as in the case of *De vita et moribus* – Gassendi's decision to publish was made under some pressure from friends and from his publisher (*Animadversiones* 1.102–3).[49] Reluctance to publish, however, was a recurring pattern with Gassendi.[50] Starting from roughly the time of publication of the *Animadversiones*, Gassendi began a new and somewhat more organized version of the Epicurus project that would become the *Syntagma Philosophicum*. At this point, he stopped billing the Epicurus project as a commentary and publicly acknowledged it as a presentation of his own philosophy. Although Gassendi started putting the *Syntagma* together only six years before his death, much of the material it incorporates is taken over from early writings – some dating back to the mid-1630s – without a great deal of revision. Most of the material from the MS Harley 1677 preface, for instance, is incorporated without significant changes, although a great deal is added.

Based on the accounts of Bloch and Palmerino, we have the following dating of final drafts of the various sections of the *Syntagma*:

1644: *Opera* 2.1–62 and 2.112–92 ("De Rebus Terrenis Inanimis," "De Globo ipso Telluris," "De Lapidibus, ac Metallis," "De Plantis" of *Physics* 3a = *De Rebus Terrenis Inanimis*)
1644–5: *Opera* 2.193–658 (*Physics* 3b = *De Rebus Terrenis Viventibus*)
1649: *Opera* 1.1–30 (the *Liber Proemialis*), a significantly expanded version of the *De vita et doctrina* preface completed in March 1634

49 This is further discussed in Joy, *Gassendi the Atomist*, 70–1.

50 In reply to Sorbière's entreaty to persuade Gassendi to send him anything ready for publication, for instance, Hobbes pointed out with some irritation that "if he is not moved by your letters, he will be even less moved by my words" (October 22, 1646). Hobbes, *The Correspondence*, 2.142.

1649–55: *Opera* 1.31–699 (the *Logic*; *Physics* 1; "De Caeli," "De varietate Siderum," "De motibus siderum," and ""De Luce Siderum" of *Physics* 2)

Thus, we have dates for all of the *Logic*; all of the *Physics* save for "De Cometis," "De Effectibus Siderum," and "De vocatis vulgo meteoris," and none for the *Ethics*. It is worth noting that the material in volume 2 of the *Syntagma*, including the account of generation to be discussed in Chapter 8, was written *before* volume 1.

It is slightly harder to determine when the *Ethics* was written because no manuscript evidence survives. Pintard and Rochot thought that the *Ethics* was finished by 1634, while Bloch dates it to 1642 at the earliest, and probably the late 40s.[51] Bloch's main evidence is that book 3 of the *Ethics* – "On liberty, fortune, fate and divination" – seems to be a revision and expansion of material on fortune and liberty from the *De vita et doctrina* manuscript dated to 1641. By the time of the *Animadversiones*, this material had been shifted from *Physics* to *Ethics*. Sarasohn concurs with Bloch on this dating and argues that the reason the *Ethics* appeared so late is that it was stimulated by Gassendi's interactions with Hobbes after Hobbes's 1640 return to Paris.[52]

The *Syntagma* forms volumes 1 and 2 of the *Opera Omnia*. The third volume is devoted to "Opuscula Philosophiae," the fourth to "Opera Astronomica," the fifth to "Opera Humaniora ac Miscellanea," and the sixth to a selection of Gassendi's Latin correspondence. French correspondence is left out entirely, although the French correspondence between Gassendi and Peiresc is available elsewhere.[53] Rochot describes a manuscript where Gassendi wrote out the order in which he wanted his *oeuvre* to appear.[54] Hence we know, for instance, that Gassendi wanted the second book of the *Exercitationes* published posthumously – and that he did *not* want the whole of the *Animadversiones* published but only the text, translation, and philological notes. This reinforces my decision to focus on the posthumous *Syntagma* rather than on the work published within Gassendi's lifetime.

Another reason for preferring the *Syntagma* to the *Animadversiones* is that it is less dependent on Epicurean doctrine. The *Animadversiones* is

51 Pintard, *La Mothe le Vayer–Gassendi–Guy Patin*, 35.

52 Sarasohn, *Gassendi's Ethics*, 125 ff.

53 Tamizey de Larroque, *Lettres de Peiresc*, 4.177–611. Gassendi, *Lettres latines*, includes more of Gassendi's Latin correspondence.

54 Rochot, "Vie et Caractère," 46–7.

primarily a line-by-line commentary, although Gassendi is careful to note those points on which Epicurus cannot be followed or approved – in particular, those having to do with God, providence, and the human soul. The *Syntagma* incorporates a great deal of material from the *Animadversiones,* but Gassendi is careful to point out to the reader that he is not an Epicurean, indeed, that he adheres to no sect but rather approves doctrines of one sect or another according to what seems most probable (1.29b). He notes that he might seem to favor Epicurus – especially because he has tried to clear Epicurus' name – but this is simply because the Epicurean mode of explanation on the basis of atoms and the void in physics, and on the basis of pleasure in ethics, seems the most useful to him (1.30a). This does not mean, Gassendi adds, that he approves all of Epicurus' ideas on religion, nor that he is so attached to any of Epicurus' ideas that he takes them for certain (Ibid.).

This note is significant for two reasons. It is intended to absolve Gassendi from possible misinterpretations arising from the connection commonly believed to exist between Epicureanism and atheism or libertinism. It is also a declaration of genuine philosophical independence – the sort of philosophical independence, indeed, that the *Exercitationes* attacks scholastic Aristotelians for lacking.

Galileo and the Question of Copernicanism

In the final section of this chapter, I turn to one of Gassendi's most significant early influences, Galileo, and the way in which Gassendi dealt with the condemnation of Galileo.[55] Although Gassendi was generally careful to avoid explicit commitment to Copernicanism, the widely read and highly controversial *De motu* of 1642 is, as Morin noted, clearly intended as a defense of Copernicanism. Moreover, Gassendi's account of the motion of macroscopic bodies is heavily reliant on the Galilean theory of motion that was inseparable from his Copernicanism. Thus, the condemnation of Galileo was a crucial event in Gassendi's intellectual life.

Gassendi adopted the Copernican hypothesis quite early on in his career. The published preface to the *Exercitationes,* for instance, says that in the never-written Book IV, "rest will be brought to the fixed stars and the sun: but the earth, as if one of the planets, will be made to move"

[55] For an account of the reaction of Gassendi's intellectual circle to the condemnation of Galileo, see Sarasohn, "French Reaction to the Condemnation of Galileo."

(3.102).[56] In August of 1625, Gassendi wrote to Galileo, sending him a copy of the *Exercitationes* and conveying some of his observations on sunspots, which helped confirm Galileo's claim that they do not exhibit retrograde motion (6.4a–6a).[57] The letter makes Gassendi's admiration of both Galileo and Copernicanism quite clear: "[F]irst, my Galileo, I would like you to consider yourself assured that I embrace your Copernican opinion in Astronomy with so much mental pleasure that because of it, I seem to be liberated and wander with free mind through the immense spaces, the barriers of the common system of the world having been broken off" (6.4b). This attitude persisted throughout Gassendi's life. Some further correspondence followed (6.10a–11b, 6.36b–37a), and Gassendi also learned of Galileo's work through Peiresc, as when Galileo sent Peiresc a copy of the *Starry Messenger* along with a telescope for Gassendi.[58] In 1632, the *Dialogue Concerning the Two Chief World Systems* was published, much – as Gassendi tells us – to Peiresc's joy (*Mirrour* 4.27). That March, Gassendi sent Galileo a copy of *Mercurius in sole visus*, the record of his observation of Kepler's predicted 1631 transit of Mercury (6.45b). In November, he wrote again, saying that he had received a copy of the *Dialogue* and praising both the book and Galileo's genius (6.53b–4a).

On April 30, 1633, Gassendi's astronomer friend Ismail Bouillau informed him that Galileo was in Rome, where he was to respond to the objections of the Inquisition (6.412a–b). Bouillau found it hard to believe that Galileo was guilty of anything, and seems to have assumed that Gassendi would feel the same way. Indeed, a few weeks later Gassendi wrote to Tommaso Campanella that "from ample recent letters from Galileo I have found out that he will soon be in Rome, whence he has been summoned. This is surprising, since he has published nothing without approval; but it is not for us to know about these momentous things" (6.56b). Both Gassendi's sympathy and his caution are evident.

It is unclear exactly when news of Galileo's June 22 condemnation and abjuration reached Gassendi. Although Peiresc's letters to Gassendi during this period are full of gossip about people, coins, and telescopes, they do not mention Galileo. By August 3, Gassendi had received from Diodati a copy of a letter Galileo sent, but he does not say what precisely was in

[56] For further evidence, see Brundell, *Pierre Gassendi*, 30 ff.

[57] Since all the known planets exhibited retrograde motion, this undercut Jean Tarde's theory that sunspots are in fact tiny planets closely circling the sun. Baumgartner, "Galileo's French Correspondents," 175.

[58] Ibid.

that letter. Indeed, for some time, Gassendi seems to have been unsure of exactly what was going on. Early in 1634 he wrote to Galileo again: "I hold fast in awaiting [news of] what has happened with you. . . . For although ignorant rumor is spread persistently, I hardly trust it until the matter has been seen clearly" (6.66b). He recommends serenity, but the letter is not entirely disinterested. Gassendi also hopes that Galileo might send him a new telescope lens, since none as good, he says, can be found in Venice, Paris, or Amsterdam. He continued friendly correspondence with Galileo. When Galileo lost sight in one eye Gassendi tried to comfort him with his notorious claim that we can only see out of one eye at a time anyhow or, in case that failed to console, with remarks on his own sadness at the recent death of Peiresc (6.94a–5a). None of the surviving correspondence mentions Copernicanism or Galileo's condemnation. And although both Peiresc and Gabriel Naudé expressed the suspicion that Christoph Scheiner (with whom Galileo had disagreed about sunspots) was at the root of the condemnation, Gassendi never echoes this opinion, and maintained friendly correspondence with Scheiner afterwards.[59]

Gassendi's Paris circle was interested not merely in Galileo's Copernicanism but also in his sciences of motion and mechanics. Mersenne's *Harmonie Universelle*, for instance, described the science of motion as presented in the *Two World Systems*. His 1634 pamphlet *Traité des mouvemens, et de la cheute des corps pesans* described the measurements that convinced him to accept Galileo's odd-number rule for the proportion of acceleration.[60] And when the *Two New Sciences* was published in Holland in 1638 after having been smuggled out of Italy, Mersenne very quickly published a free and partial translation, *Les nouvelles pensées de Galilée*. Mersenne's involvement assured the spread of knowledge to people like Gassendi, Bouillau, Roberval, Descartes, and Huygens.[61]

Gassendi first discusses Copernicanism in the three letters *De motu*, letters whose subsequent publication occasioned a protracted and acrimonious dispute with Jean-Baptiste Morin. *De motu* takes as its starting point the accounts of uniform motion, naturally accelerated motion,

59 See Gorman, "A Matter of Faith?" 285.

60 The odd-number rule says that a body in free fall that falls through one unit of space in the first unit of time will fall through three units in the second time, five in the third, and so on. This is, of course, equivalent to saying that total distance fallen is proportional to the square of time in free fall. For an account of Mersenne's works on motion, see Palmerino, "Infinite Degrees of Speed."

61 Galluzzi, "Gassendi and l'Affaire Galilée," 510.

and projectile motion in the Third and Fourth Days of Galileo's *Two New Sciences* and attempts to explain the underlying physical causes of such motion. In particular, Gassendi's *De motu* attempts to divine the physical causes for the rule for acceleration first proposed in the *Dialogue*. However, *De motu* is also quite clearly intended to defend what Gassendi calls the "Galilean theorem" that *If the body we are on is in motion, everything we do and all the things we move will actually take place, and appear to take place, as if it were at rest.*[62] Both Gassendi and those who objected to his *De motu* believed that if this "theorem" is true, a common antiheliocentrist objection – that bodies on a moving earth would not behave as they actually do – would be defeated. Thus, Gassendi's *De motu* attempts to provide the integration of the *Dialogue* and *Two New Sciences* that Galileo was prevented from carrying out by his condemnation.[63]

Because observing terrestrial phenomena is no help, Gassendi claims that the judgment whether heliocentrism or geocentrism is more probable will stand or fall on careful celestial observation. Such observations provide several pieces of evidence that the apparent motions of the heavens probably result from the earth's motion and not genuine heavenly motion. For instance, the apparent occasional standstill and retrograde motion of Mercury, Venus, Mars, Jupiter, and Saturn is better explained by heliocentrism than geocentrism (3.515bff).[64] Moreover, the hypothesis that the earth moves is much simpler than the hypothesis that the heavens move around it (3.506a–b). It is interesting that Gassendi seems to think the greater simplicity of the heliocentric hypothesis is evidence for its physical reality as well as its superiority qua hypothesis.

A commitment to Copernicanism seems pretty clear from all this, but Gassendi concludes with a cautious reminder to his readers:

> I set about only to speak of the stupidity of that argument from the motion of projectiles commonly provided for the rest of the earth. . . . Do not require me to repeat that I did this not to assert the motion of the earth but from a love of truth,

[62] Galileo never calls this a theorem.

[63] Galluzzi, "Gassendi and l'Affaire Galilée," 511–12.

[64] Here Gassendi is arguing against the epicycles typically used to explain retrograde motion. One tactic he uses is to point out that according to geocentrism, the earth behaves differently from all the other planets. This, Gassendi suggests, is at best strange since the earth is one planet among many. Gassendi also presents a version of Galileo's explanation of the tides in terms of the motion of the earth and the gravitational attraction of the moon. See Palmerino, "Gassendi's Reinterpretation of the Galilean Theory of Tides."

to suggest that its rest should be established on firmer reasoning. Also, do not require that I communicate [such firmer reasonings] to you, as if you supposed that I had some; indeed, if I did, I would [communicate them] willingly now. (3.519a)

Instead, Gassendi notes two reasons against asserting that the earth moves – the 1633 decree and the existence of Biblical texts apparently implying heavenly motion and terrestrial rest.

A standard Galilean strategy for dealing with the apparent conflict between Copernicanism and such Biblical texts was to interpret the relevant passages as concerning apparent, rather than real, motion. Gassendi's comment on this strategy is noncommittal:

[A]lthough the Copernicans maintain that the passages of sacred Scripture that attribute stationariness or rest to the earth and motion to the sun should be explained, as they say, as about appearances and in regard to our capacities and common ways of speaking.... Nevertheless, because these passages are explained differently by men who are agreed to be of great ecclesiastical authority, I stand apart from [the Copernicans] and on this occasion do not blush to make my intellect captive. This is not because I judge that it is an article of faith; for (as far as I know) that assertion is separate from those that are either promulgated or received according to the Universal Church: but because the judgment of [the Copernicans] should be considered a prejudice which cannot be of great moment according to the faithful. (3.519b)

Throughout his work, Gassendi is tempted by but not committed to the Galilean program of interpreting scripture allegorically, based on natural knowledge. In the *Syntagma*'s account of creation *ex nihilo*, for instance, he is careful to note that several Church Fathers held that the Biblical claim of a six-day creation should be understood as allegorical, since God's power is such that he could have done so in a moment.[65] However, he does not go so far as to endorse the view, writing instead that we cannot fall into error by construing the words of Genesis "as they sound" (1.485a). The second decree of the fourth session of the Council of Trent had stated that

in matters of faith and morals pertaining to the edification of Christian doctrine, no one, relying on his own judgment and distorting the Sacred Scriptures according to his own conceptions, shall dare to interpret them contrary to that sense

[65] The situation here is complicated. For Gassendi, as was standard, articles of faith can be provided by the writings of the Church Fathers as well as by the Bible, the Pope, and the decrees of the various Councils. Thus here he is presenting the situation as one in which the Bible, literally read, says one thing and the Fathers another.

which Holy Mother Church, to whom it belongs to judge their true sense and meaning, has held and does hold, or even contrary to the unanimous agreement of the Fathers.[66]

Gassendi endorses this, as was necessary, and his tendency in theological matters to focus on Patristic texts over those of scholastic theology is entirely consonant with it. However, what counts as an article of faith under Trent was much disputed, and one hard-line position, put forward most notably by Robert Bellarmine, held that "not only the opinions in the Bible but also all the words individually pertain to faith."[67] Gassendi could not and did not accept this broad reading. Well into the 1640s, he insisted that geocentrism is not a matter of faith, and thus that the relevant passages of the Bible do not require a literal reading.

Let us return to the *De motu* passage where Gassendi refers to the Copernicans' judgment as a "prejudice." He neither makes clear why it should be considered a prejudice nor whether the prejudice is holding that the earth moves or, rather, reading the relevant Biblical passages as concerning merely apparent motion. There is no technical definition of a prejudice (*praejudicium*) common to Gassendi's various works. However, one way in which he might think that heliocentrism is a prejudice is that it goes beyond the appearances and what they univocally signify. Thus, Gassendi could deny that we are justified in ascribing physical reality to *either* Copernicanism or Tychonism.

Such a tactic has Epicurean credentials. The Epicureans thought one should treat cases where only one underlying state of affairs explains the appearances, and cases where there are multiple and incompatible underlying states of affairs competing to explain the appearances, differently. Indeed, in the *Syntagma* – although not in *De motu* itself – Gassendi applies the Epicurean treatment of hypotheses to astronomy explicitly (1.630a). However, it is worth noting that many sixteenth-century astronomers had refrained from ascribing physical reality to their hypotheses for entirely non-Epicurean reasons.[68]

One source of concern regarding Copernicanism was Biblical literalism; another, perhaps more worrying, was the 1633 decree. Gassendi

[66] Blackwell, *Galileo, Bellarmine, and the Bible*, 183. Translation of the Decrees of the Council of Trent Session IV (8 April 1546), "Decree on Tradition and on the Canon of Sacred Scripture."

[67] Ibid., 31, quoting Robert Bellarmine's *Disputationes de controversiis Christianae Fidei adversus hujus temporis hereticos*, 2.2.12.

[68] Jardine, *The Birth of History and Philosophy of Science*, 72–82.

is carefully vague about what, precisely, was decreed, and in particular whether the decree held geocentrism to be "a matter of faith and morals." This vagueness may initially have been a matter of inadequate information, but it is plausible that Gassendi later deliberately misrepresents his knowledge of the condemnation. Consider his account of some famous objections to heliocentrism in the second letter *De motu*, which ends, after claiming that all these objections are seriously flawed, with the statement: "I am in the situation of revering that decree in which some Cardinals are said to have approved the rest of the Earth" (*De motu* 2.13; 3.519a). The phrasing is noteworthy; even in the 1640s, Gassendi is careful to make clear both that he has merely heard reports of the decree, and that the decree was handed down by cardinals rather than the Pope himself.[69]

The "seriously flawed" objections to heliocentrism of *De motu* were, for the most part, those of Jean-Baptiste Morin, and the first two letters *De motu* instigated a polemic with Gassendi's old friend and fellow astronomer.[70] (The Jesuit Pierre Le Cazré had also written objections, although somewhat more friendly ones.[71]) Morin, who had already written three anti-Copernican works, produced a fourth, *Alae telluris fractae* (*Breaking the Wings of the Earth*) in response to Gassendi. *Alae telluris*, like Morin's previous anti-Copernican texts, relies mainly on the argument that the behavior of moving bodies on the earth disproves Copernicanism (*Receuil* 105). Morin rightly suggests that Gassendi's claim that he was not arguing *for* heliocentrism, but merely *against* bad geocentrist arguments, was disingenuous. He also contests Gassendi's claim that "several cardinals" and not the Pope himself had promulgated the decree. Gassendi replied to Morin in the third letter *De motu*, where he repeats that he is not committed to Copernicanism: His goal is to defend an account of motion that works whether or not the earth moves. He merely wishes to point out that, on his account of motion, the claim that the behavior of falling bodies disproves heliocentrism fails. Further, Gassendi argues that his speculations about gravitational attraction are independent of Copernicanism: One could, like Athanasius Kircher, construe gravity as extrinsic to heavy bodies while being a committed geocentrist (3.561b–3b).

69 It is generally agreed that Urban VII wrote the decree himself.

70 In the letters of the *Receuil*, both Morin and Gassendi refer to their "ancien amitié" repeatedly.

71 For more on Cazré's objections, as well as Morin's, see Galluzzi, "Gassendi and l'Affaire Galilée."

Further exchanges of public letters followed, and Morin's focus shifted from heliocentrism to Gassendi's dismissive arguments against judicial astrology, which Morin sees as not taking sufficiently into account his own massive *Astrologia Gallica.* Some of these letters were printed by Gassendi's friends in 1650, together with a rather extravagant preface referring to Morin's "libel" and describing Gassendi as "following the example of Jesus Christ" and turning the other cheek.[72] Unsurprisingly, these claims are not consistent with Morin's account, and indeed Morin claims that Gassendi was slandering his name in 1646 in Provence (rather than the other way around). The polemic with Morin continued: Morin attacked Gassendi's atomism directly in his *Réponse à une longue lettre de M. Gassendi* (1651) and *Defensio suae dissertationis.* Two replies, titled *Anatomia ridiculi muris* and *Favilla ridiculi muris* (*The anatomy of a ridiculous mouse, The ashes of a ridiculous mouse*[73]) were published under François Bernier's name, although Sorbière says that Gassendi wrote them himself.[74] However, Gassendi's problems with Morin do not appear to have had any further ramifications, and it is worth noting that the *De motu* dispute is the only occasion on which Gassendi faced serious charges of heterodoxy.

Perhaps because of the Morin controversy, Gassendi's later discussions of Copernicanism are somewhat more cautious. In the *Syntagma,* just as in *De motu,* Gassendi is unwilling either to state that he accepts heliocentrism or to grant that geocentrism is a matter of faith:

> there are three principal systems by which the order of the world can be conceived, but which of the three should be preferred to the other two is a great question. Now this cannot be determined except from deductions concerning celestial motions. . . . It appears that the common or Ptolemaic system is least probable, on account of many things, principally that it is now established that Venus and Mars are not always on the same side of the sun. . . . So, since it appears to be necessary that one of the other two remaining systems be judged to be preferable, it appears that the Copernican system is clearer and more elegant; but because there are sacred texts that attribute rest to the earth and motion to the sun, and it is reported that a decree exists commanding that those texts be understood as not about apparent but about genuine rest and motion, thus it suffices that the Tychonic system should be approved and defended by those who revere such a decree. (1.149a)

[72] *Receuil,* preface, first two unnumbered pages.

[73] The reference is to a line of Horace, "Parturient montes, nascetur ridiculus mus" (those who wish to give birth to mountains will produce a ridiculous mouse). *Ars Poetica* 139. I owe the reference to Fisher, "Science and Skepticism in the 17th Century," 171.

[74] Sorbière to Hobbes: Hobbes, *The Correspondence,* 2.388.

Although one assumes that this is merely the stance of caution, it is worth remembering that it is entirely in keeping with Gassendi's Epicurean methodology to simply suspend judgment in such a case.[75]

The *Institutio astronomica*, written as lectures when Gassendi was professor of mathematics at the Collège Royal, does not put in as much effort to suggest Copernicanism while remaining overtly neutral. Gassendi presents the Tychonic system, the Copernican system, and the responses of the Copernicans to objections from astronomy, physics, and Scripture. The last are the most interesting for current purposes. As well as repeating the previously seen observations that Biblical talk of the earth's fixity may simply be talk for the vulgar or that "fixity" may mean "incorruptibility" (3.519b), Gassendi repeats an elegant piece of rhetoric from *De proportione*. The fact that certain Biblical interpreters and doctors of religion think heliocentrism conflicts with Scripture does not show that heliocentrism is incompatible with the articles of faith, any more than the fact that some interpreters think the Bible contradicts the Aristotelian theory that the earth is surrounded by spheres of air, fire, and ether showed that those Aristotelian views were contrary to faith (3.641b).

Gassendi also introduces a new argument in the *Institutio astronomica*, claiming that the condemnation of Galileo need not be taken to apply to Copernicanism in general:

> [T]he opinion commanded by the Congregation of Cardinals of the Inquisition is customarily relied on here, which condemned Galileo's opinion about the motion of the earth: But the orthodox respond... that this opinion was special, or pertained only to Galileo, inasmuch as there could be special reasons against [Galileo's view] that were not valid against others. They add that this opinion should indeed be weighed as of vast moment; but nevertheless it need not on that account be taken as an article of faith, in the way that those things decreed by the General Council are considered. (4.60a–b)

When Gassendi concludes by saying that Tychonism is to be preferred to Copernicanism on the grounds of "probability," this comes somewhat unexpectedly. It is easy to read the *Institutio astronomica* as Copernican in spite of this; Olivier Bloch, for instance, writes that Gassendi's refusal to endorse Copernicanism in his own name is "the only sign of prudence" evident.[76] For Gassendi never seems willing to recognize what

[75] For an argument that Gassendi's account of the reality of hypotheses came about because he needed to find an intellectually acceptable way to respect the condemnation of Galileo, see Brundell, *Pierre Gassendi*, 43 ff.

[76] Bloch, *La philosophie de Gassendi*, 329.

is commonly taken to be the central point of the 1633 condemnation, that Copernicanism is an article of faith and not something to which the freedom of philosophizing pertains, *regardless* of the sort of philosophical reasons advanced.

I bring up Bloch's diagnosis because Bloch understands Gassendi's attitude toward Copernicanism as evidence for his thesis that Gassendi's philosophy is implicitly materialist and atheist but that Gassendi himself refused to allow these views to surface. Bloch takes the doctrine of double truth – on which reason leads to one set of truths and faith to another, incompatible set – to be Gassendi's defense against the possible charge of disingenuousness or duplicity in his philosophical system.[77] Although his diagnosis of the Copernicanism case is entirely plausible, I suggest that the larger thesis relies on significantly overestimating the tension between Gassendi's natural-philosophical claims and the relevant articles of faith. Far from being forced to rely on the doctrine of double truth, Gassendi is in fact very careful to prevent conflict and inconsistency. And although he is not entirely successful in the Galileo case, this has more to do with peculiarities of the Church's position during the Galileo affair than with deep-seated tensions within Gassendi's own account. I shall return to this point in Chapter 10, after we come to understand the scope and epistemic status of Gassendi's natural-philosophical claims.

[77] Ibid., 328.

2

Gassendi's Philosophical Opponents

Gassendi opposes the three main intellectual systems of his day: scholastic Aristotelianism, Renaissance neo-Platonism, and the philosophy of Descartes. His critiques of Aristotelianism and neo-Platonism are among his first published works, and his *Objections* to the *Meditations* were written during the relatively early stages of composition of the *Syntagma*: Gassendi presented his opposition to the major philosophical schools of his day before his positive views were fully developed.

Gassendi draws both a rhetorical and a methodological distinction between clear and obscure philosophy and suggests that all three intellectual systems contain elements of obscurity that render them unacceptable (1.13b–17a). The foremost representative of the clear philosophy is, as one might expect, Epicurus. Gassendi acknowledges that there are points where Epicurus is less than clear – as in his accounts of the gods and the nature of the human soul – but it is now possible to give a revised, clear version of Epicurus' philosophy. (It is worth noting that the Epicurean accounts of the gods and the human soul are not rhetorically or methodologically particularly different from the rest of the Epicurean system, but they are strikingly different in terms of their acceptability to a Christian audience.) Other clear philosophers include Pyrrho and the Stoics. Empedocles, Parmenides, Pythagoras, Plato, Heraclitus, and Aristotle exemplify obscure philosophy.[1] Gassendi's favorite contemporary, Bacon,

[1] Gassendi is careful to limit this to what he calls the exoteric philosophy of Aristotle, while making no attempt to comment on the content of his esoteric philosophy. We now know that the surviving works of Aristotle are the esoteric philosophy, while the exoteric philosophy has been lost.

is clear, and Robert Fludd, for instance, is obscure. The stated criteria for obscurity are rhetorical. Obscure philosophers express their doctrines in *fabulae* or write in verse; they use "symbols" and "enigmas" to explicate their meaning; and they affect obscure terminology – this last being the crime Aristotle is particularly guilty of. Gassendi has methodological worries about couching a philosophy in symbolic terms. The clear philosophy, he tells us – at least in the best case, the Epicurean philosophy – is written without reliance on symbols or stories and can be propounded without reliance on logic. Indeed, Gassendi's early objections to scholastic Aristotelianism, presented in his first published work, the *Exercitationes*, are primarily objections to Aristotelian logic.

None of the great diversity of views within late scholasticism is apparent from Gassendi's treatment of Aristotelianism, either in the *Exercitationes* or in his later work. He almost never identifies individual scholastics or discusses their disagreements. His attack on Aristotelianism is an attack on an oversimplified position of the sort presented in textbooks he used in teaching like Eustachius' *Summa.*[2] However, it is useful to examine the *Exercitationes* because of the clarity with which Gassendi there states themes he will maintain throughout his career.

We are now thoroughly accustomed to thinking of the "new philosophers" of the seventeenth century as both reacting against and borrowing from a complex, differentiated Aristotelianism. However, we are still beginning to see how they both react against and borrow from other strands in late Renaissance and early modern intellectual culture as well. Gassendi is far more directly engaged with neo-Platonism than with university-based Aristotelianism. This may, in part, be a result of his Provençal education. Aix, near the Italian border, had as much contact with Florence and Pisa as with Paris. Yet it also reflects more generally the intellectual culture in which seventeenth-century philosophers found themselves.[3]

Gassendi is much more explicit about the diversity among neo-Platonists than among Aristotelians. Indeed, even though he groups "the doctors" and "the interpreters of Aristotle" together en masse, there is no talk of a group of neo-Platonists as such, but merely of various figures

[2] Eustachius, *Summa philosophiae quadripartita.* The *Exercitationes* often quotes Eustachius: See, for instance, *Exercitationes* 1.1.14 (3.110b). For Gassendi's pedagogical use of Eustachius, see Armogathe, "L'Enseignement de Pierre Gassendi."

[3] Leibniz, in a letter to Thomasius of 1669, recognizes three chief parties in philosophy: the Aristotelians; the followers of Paracelsus and Van Helmont; and the Cartesians. Leibniz, *Philosophical Papers and Letters,* 96.

who accept a version of the doctrine of the *anima mundi.* I group these figures together as neo-Platonists simply for convenience. For Gassendi is concerned throughout the *Syntagma,* as well as in some occasional works, to refute the doctrine of the world soul in general.

Descartes was not a formative influence on Gassendi in the way Aristotelianism and neo-Platonism were. He had published nothing when Gassendi began working on his Epicurus project. However, it is instructive to see how Gassendi confronts a rival intellectual system that we understand relatively well.

Scholastic Aristotelianism

In 1624, Gassendi published the *Exercitationes Paradoxicae Adversus Aristoteleos,* ostensibly concerning Aristotelian doctrine in general.[4] A second volume was written shortly thereafter but published only posthumously. The arguments of the *Exercitationes* are both philosophical and philological, the latter sort focusing on how accurate the extant texts are and how well they have been interpreted. Gassendi argues that the Aristotelian corpus has not received the careful reconstruction and interpretation that philosophical use of it requires. His detailed commentary on the surviving Epicurean texts in the *Animadversiones* will later provide a model of the sort of philological and interpretive work he takes to be necessary.

In his preface, Gassendi tells us that the complete *Exercitationes* will include books entitled *Physics, Simple Bodies, Mixed Bodies, Metaphysics,* and *Ethics* as well as the two on logic and Aristotelian philosophy in general that he actually wrote. In the *Physics,* he says, he will demonstrate that natural movement does not have the source it is commonly thought to have; recover the space of the ancients to replace Aristotelian space; and introduce the void and a new theory of time. The *Syntagma* later carries out all these plans. Gassendi describes *Simple Bodies* as "attributing rest to the fixed stars and motion to the earth," as well as arguing against the Aristotelian view of the elements. This Copernican project was revised after the 1633 condemnation of Galileo. In *Mixed Bodies,* Gassendi plans to discuss comets, the passage of the chyle through the stomach, the fact that seeds are animate, and more familiar issues like the reasoning ability of brutes and the identity of intellect and imagination. In *Metaphysics,*

[4] They are "paradoxical" exercises because they are intended to show that Aristotelian doctrines ultimately lead to paradox or internal contradiction.

he plans to argue that "all our cognition of God and angels depends on faith," so that "arguments about the nature and operations of separated substances are entirely in vain" (3.102). This argument is more or less abandoned by the time of the *Syntagma* in favor of reliance on the argument from design and on a somewhat more sophisticated account of cognition of imperceptible entities. Finally, *Ethics* will defend the Epicurean doctrine that pleasure is the highest good – a view that Gassendi will maintain, with the caveat that the highest pleasure is the beatific vision, throughout his work.

The two extant books have three chief targets, which Gassendi continued to oppose throughout his career. His objections to Aristotelianism formed the starting point for the logic and epistemology developed in the *Syntagma.* The first of Gassendi's three targets is the scholastic Aristotelian conception of logic, which he attacks on the familiar grounds that it simply generates verbiage without leading to any new discoveries. The second target is the Aristotelian doctrine of *scientia.* Although he allows the possibility of knowledge – even, later, knowledge of unobservable essences – none of this counts as *scientia.* Gassendi's third target is the set of fundamental metaphysical and logical entities of Aristotelian natural philosophy. He denies that universals exist or that there is any distinct intellectual faculty that operates over them and argues that the ten categories are groundless reifications of the way we conceive the world. Gassendi also mounts an attack on hylemorphism, and I shall consider this in some detail later when considering his search for the efficient principle of the material world.

It was common in the seventeenth century to complain that scholastic logic is useless as a method of discovery, although it is now recognized that Aristotelian logic was never intended as a method for discovering new truths but rather as a method of justification and teaching. Indeed, this, along with many of Gassendi's other complaints, is found even in Patrizi's 1581 *Discussionum Peripateticorum,* which is often thought to have influenced the *Exercitationes.*[5] Gassendi goes along with this common objection, rejecting definition in terms of genus and difference as futile on the grounds that no definitions will help us understand the nature of the sun or even give us "complete knowledge of the nature of a flea" (3.150a). For when someone makes a mistake, "you will discover that the cause is

[5] This book was available to Gassendi in Peiresc's library. Bloch, *La philosophie de Gassendi,* 112; Joy, *Gassendi the Atomist,* 240; Gassendi, *Dissertations en forme de paradoxes,* xi (editor's introduction).

not that he does not know logic but that he has not examined the thing enough" (3.151b–3a). Only observation can help us come to know the natures of things, but even observation does not give full knowledge of natures. Similarly, Gassendi objects that logic cannot keep its promise to differentiate truth from falsehood. Rather, he says – using the examples of the proportions between the lengths of the strings that emit the various notes, and the properties of geometrical figures – that "just as each science has its own truth to be learned, each science has its own measures for knowing it." In geometry, the quadrant is the measure of truth; in arithmetic, calculation; in physics, the senses; and so on (3.151b–3a). This leaves no room for purely logical criteria.

Gassendi further argues that traditional logic is useless even for pedagogical purposes, claiming that the ability to teach stems not from knowing logic but from knowing the subject: "is there anyone who wouldn't prefer to have navigation explained by someone who has sailed a certain coast for a long time" rather than by a logician (3.150b–1a)? The dialectician will reply that the analogy is unfair because logic is concerned with the teaching of sciences and navigation is not a science. But Gassendi seems to be making no real effort to be fair in his attacks. The arguments of the *Exercitationes* are not directed against particular philosophers but against something like a very simple textbook. The point of Gassendi's diatribe against Aristotelian logic is thus more to take sides in an ongoing quarrel than to make philosophical progress; Gassendi is aligning himself with modern, humanist writers against the scholastics. Thus, it is not the substance of Gassendi's complaints against Aristotelian logic that are important for our purposes but merely the fact that he makes them in his first published work.[6]

Gassendi's denial of the possibility of *scientia* is well known from Richard Popkin's discussion of him as a mitigated skeptic, that is, someone who holds that there is no *scientia* but that we have knowledge of another sort within certain restricted domains.[7] It will turn out, however, that the skeptical and probabilistic themes in Gassendi's epistemology emerge at different stages in his career. Even though the theory of signs, which will be discussed in Chapter 4, forms the basis of Gassendi's account

[6] The reader may wonder whether Epicurean logic is liable to the same complaints: Can it serve as a method of discovery? Or is it pedagogically valuable? However, Gassendi had not yet chosen Epicureanism as a replacement for Aristotelianism when he made these complaints. Moreover, logic plays a comparatively small role in Epicureanism, and this may initially have been appealing to Gassendi.

[7] Popkin, *The History of Scepticism*, 129 ff.

of probable knowledge, his rejection of *scientia* occurs very early on in the *Exercitationes* itself.

The doctrine of *scientia* has three main components: a progression from causes to effects; a progression from the universal to the particular; and the use of essential definitions. Gassendi attacks these three components in turn, beginning with the claims that knowledge should proceed from causes to effects and from the universal to the particular:

> An a priori demonstration is from causes and the more universal, an a posteriori demonstration from effects and the less universal. But aren't effects better known than causes because investigating causes follows observing effects? And aren't particular or less universal things better known than more universal because we get the more universal by induction from the previously known less universal? Thus, an a posteriori demonstration is better known or proceeds from the better known than an a priori demonstration. Hence it is more certain, or proceeds from the more certain. For certainty is by greater or more evident knowledge. (3.191b)

As we shall see, Gassendi's account of knowledge through signs inverts the order of knowledge, moving from an effect to its necessary cause. Gassendi's claim is, in effect, that nothing further is added to our knowledge by deducing the effect from its cause; once we have arrived at cognition of the cause, our knowledge is as complete as it can be. It is worth noting that the objection applies not only to a very simplified account of *scientia* but also to *regressus* theory (3.191b). For by rejecting any distinction between what is better known to us and what is better known by nature, Gassendi rules out the usefulness of the second step of the *regressus*.

The final target in Gassendi's attack on *scientia* is the claim that there are universal essences of things that can give content to our definitions and serve as useful first principles. For Gassendi, there are no universals beyond human thought (or, in less careful moments, human language):

> [U]niversals are nothing more than what grammarians call common nouns... individuals are nothing more than proper nouns. What, you say, do you then accept the crazy opinion of the nominalists, who recognize no universality other than [the universality] of concepts or names? This is so... nothing can be found which is not a particular thing. Then where will you point out that this tribe of universals lives? (3.159a)

Gassendi's reference to "the crazy opinion of the nominalists," at a time when nominalism was widespread, suggests that the argument is primarily rhetorical. He continues poking fun at universals, asking "if someone sliced with a sword between you and me, wouldn't he cut this common quality [i.e., the universal]?" (3.159a). The analogy is absurd and unfair: No one would ever have held that universals are bodies or, for that matter, that we should be able to point to them as the preceding passage requires. Porphyry's view that universals are *in* bodies does not imply that universals *are* bodies. Again, fairness to the opponent seems hardly the point here.

Gassendi then goes on to provide an alternate reading of statements allegedly appealing to universals, relying on the basic notion of similarity:

> when it is said that all men are of the same nature . . . this cannot have any meaning except that they are of a similar nature or that they have a similar nature, or more precisely that they have natures that are similar to each other . . . [each nature] can be conceived through one and the same concept because of this similarity. (3.160a)

Responding to the stock objection that if there are only individual natures, "the statement 'Plato is a man' will be a futile tautology," Gassendi replies that "every proposition, if it is true, must be an identity because clearly nothing should be predicated of a thing unless it is that very thing or is in that thing" (3.160b). In any true statement, the ideas must agree, and thus they must either be identical or be related as part to whole.

For similar reasons, Gassendi denies the eternality of essences and the existence of eternal truths. Unfortunately, he defers the argument against essences and eternal truths to the never-written volume on metaphysics (3.178b). However, given the language used later, in the *Disquisitio,* one can guess that Gassendi would have made the Suárezian point that the eternal truth of essential propositions consists solely in the truth of certain conditional statements. For instance, in the *Disquisitio* Gassendi argues that claims such as "man is a rational animal" are true only under the interpretation "if man exists, he is a rational animal" (3.375a). From this point on, essences in the Aristotelian sense play no role in Gassendi's philosophy. By the time of the *Syntagma,* essences are interior corpuscularian structures.

Gassendi then attacks the ten categories conceived "as sorts of reality"; that is, he attacks a reified version of the categories that takes them to be more than simply useful ways of conceptualizing things. Gassendi begins by arguing that it is arbitrary how many categories there are and how they are distinguished. One could just as well say that there is substance and what is predicated of it, stopping at two categories; or make continuous and discrete quantity separate categories, thus having more than ten (3.166b–7b). He grants that the categories make better sense if viewed "not as groups or classes of things but only [as groups or classes] of concepts." Even then, there are no such distinct and privileged groups of concepts. For instance, accidents cannot be conceived without at the same time conceiving substance. Hence, we must be careful to recognize that the categories are not properly grounded in reality (3.169a–70a).

Gassendi finds similar problems of reification in hylemorphism. In the *Exercitationes*, he has relatively little to say about form. He grants that things have forms insofar as they have shapes or patterns, but such talk should not be reified: Shapes are mere modes of matter. In later texts, Gassendi is happy to use this attenuated notion of form as long as he has something to say about what constitutes form. For instance, he says that certain crystals are hexagonal in form because they are made up of hexagonal atoms, thus explaining the structure and regularity that were thought to require a form in terms of something material.

However, Gassendi's overarching argument against hylemorphism is that forms cannot constitute the active principle within nature, that in virtue of which the natural world has causal efficacy. This is more sophisticated than the relatively superficial objections adumbrated previously, and it is the main argument against hylemorphism in the *Syntagma*. The argument occurs in Gassendi's discussion of the material and efficient principles within nature, a discussion that hinges on two closely related questions: what constitutes the nature of matter, and whether matter contains its own principle of activity or needs to be supplemented by an external principle of matter. Gassendi begins this discussion by noting that we need a material principle to explain the existence of material things and an active principle to explain the reality of secondary causation. Later works explain the active principle as built into the material principle through the motive power of atoms, but in the *Exercitationes* Gassendi merely makes the negative claim that forms cannot explain activity.

Gassendi does not consider occasionalism, the denial that there is genuine creaturely activity, as an alternative here. If a consequence of a view is that there is no secondary causation, he takes this as grounds to reject the view. For one thing, Gassendi notes, secondary causation is required in order that we, rather than God, are the authors of virtuous and vicious actions (2.817a). For another, he thinks that we experience the causality of created things in sense perception, endorsing Aquinas's claim that we know by sense that a body such as fire heats another.[8] Gassendi

[8] Aquinas, *Summa Contra Gentiles*, 3.69.

also suggests that we should read the words of Scripture "as they sound," and that a literal reading of the first chapter of Genesis – where "God commanded the Earth and Water to germinate and produce Plants and Animals" – shows that God has endowed creation with activity (1.493a; compare 1.487a). Finally, he holds that it would detract from God's power and greatness if he did not confer some active power on created things – a claim that again follows Aquinas (1.239a).[9] None of these claims are elaborated or defended in any detail; Gassendi is simply mentioning, at various appropriate points, what seem to be standard arguments for secondary causation.

Most of Gassendi's attention is devoted to attacking the views of *Recentiores* such as Telesio, Patrizi, and Digby and objecting to the *anima mundi* as an immaterial efficient principle for secondary causation.[10] However, he also considers a second immaterial efficient principle, form. Looking at how Gassendi argues that forms cannot be efficient principles shows us his more sophisticated objections to Aristotelianism and helps us understand why he adopts his atomist view of matter as active.

Gassendi is careful to attribute the view that forms are efficient principles to "certain Interpreters of Aristotle" rather than Aristotle himself. Such interpreters took from Aristotle the claims that forms are indivisible and that indivisibility distinguishes form from matter and interpreted those claims as implying that forms are immaterial entities distinct from matter (1.333b). Gassendi takes this interpretation to be dubious, for two reasons. First, there is insufficient textual basis for establishing a separate ontological category for form. Second, it ascribes to Aristotle a philosophically hopeless view, something to be avoided at almost any cost.

Why is the view hopeless? The worry is not – as some readers might think – that forms should be thought of as *formal* rather than efficient principles. For, even though it is true that forms are formal causes in respect of the body they inform, it is also entirely standard to say that forms are efficient causes in respect to the effects produced by the body they inform on *other* bodies.[11] This is the sense in which the form or – what is the same or at least closely connected – nature of fire is the efficient cause that heats a pot of water placed above the fire. The problem with the

9 Ibid.

10 He also examines some special possibilities for the efficient principle within celestial bodies, such as angels. We need not consider those here.

11 Sennert, *Epitome Naturalis Scientiae*, 54–5; Eustachius, *Summa philosophiae quadripartita*, 58–9; Du Moulin, *La philosophie, mise en français*, 6–7.

view of forms as efficient principles is instead the well-known problem of explaining where forms or active principles can come from. A traditional answer is that form is educed from the potentiality of matter.[12] Gassendi objects that this answer

> is mere words. For if they mean that it is educed in such a way that it is only a mode of matter like the shape of a statue into which bronze or wood is formed, then they are indeed saying something, but form will be merely passive, like the matter whose mode it is, and not at all an active principle. However, if they mean that it is some super-added entity, then they cannot say at all whence this entity exists, since the potential of matter has been put aside, nor from what source its power of acting comes, since the potential of matter is merely passive and in no way active. (1.335b)

Let me spell out the reasoning a bit. Gassendi takes it as a basic assumption of hylemorphism that matter is entirely passive: this, after all, is why we need a distinct active principle. Now, the claim is that the active principle is educed from the potentiality of matter. However, Gassendi objects, you cannot derive a genuinely active principle from the potential of a purely passive thing, "and therefore the situation always returns where all active potential is derived from merely passive potential" (1.335b). Thus, "one cannot hold that matter supplies the power of form" (1.335b). Hence one must either give up the claim that matter is purely passive or give up the claim that form is educed from matter.

The first option is a nonstarter. If you give up the claim that matter is passive, you have thereby given up hylemorphism, and you no longer need an active principle distinct from matter. The second option also fails, on Gassendi's account. For it simply brings us back to the original question: *Where does the activity of form come from?* Gassendi suggests that someone, abandoning the notion that form is educed from the potential of matter, might argue that form is educed from some other external thing. (Here we are still within the order of created things; the suggestion is not that God educes form.) Gassendi argues against this suggestion, however, because "there is nothing characteristic of form other than this inner power of acting" (1.335b). If we say that forms merely *transmit* activity without being an original *source* of activity, then forms as so described are doing no work, and the question of what the real efficient principle or

[12] Sennert, *Epitome Naturalis Scientiae*, 14; Dupleix, *La Physique*, 127; Leibniz, *Philosophical Papers and Letters*, 18; Pemble, *De origine Formarum*, passim; Suárez, *On the formal cause of substance*, 55 ff.

locus of activity is has not been answered. Activity is the prime conceptual ingredient of form, as the Aristotelians understand it.

A reader might not find this argument terribly satisfying. Why, she might ask, do forms need to get their activity from anything (except, of course, from God)? Why can't forms just have – or just be – active powers? Two different sorts of answers can be given here. First, Gassendi is not posing a new question but just questioning a traditional answer to a traditional question: It is traditional to ask where the power of form comes from. Second, Gassendi may be relying on the worries about reifying abstractions put forth in his earlier *Exercitationes* (3.159a ff). To say that forms just have – or, perhaps, just *are* – active powers by themselves, without any help from matter or from other causes save God, is to make forms into things or substances in their own right. Indeed, Gassendi writes that these interpreters of Aristotle must always end up saying that forms "are some true entity distinct from matter" (1.335a), a view he rejects outright on the grounds that it fails to explain the crucial connection between form and matter, turning forms into entities capable of existing independently.[13] Moreover, forms cannot just have (or just be) active powers unless their activity (or their existence) is connected in some way to the matter they are associated with. For Gassendi, as for many of his contemporaries, powers that are not the powers of some thing do not make sense. I can *conceive* of a power without a substance it belongs to, but only misguided reification of how we conceive the world could lead us to infer from this that there could *be* such powers. This line of thought brings Gassendi to reject the possibility that forms are the efficient principles within nature, and in the end to reject any view on which the efficient principle is something other than matter.

Neo-Platonism and the *Anima Mundi*

After the *Exercitationes*, Gassendi devoted little more energy to attacking Aristotelianism. The *Syntagma* often mentions Aristotle himself, but it almost never refers to Aristotelianism as Gassendi would have encountered it in the university curriculum. None of the more or less

[13] It is worth noting that some of Gassendi's near contemporaries embraced the view that there are substantial forms distinct from matter even though they denied that substantial forms could exist independently of matter. For a discussion of arguments for and against such a view of substantial form, see Des Chene, *Physiologia*, 76 ff.

contemporary figures discussed in the *Syntagma* – writers such as Digby, Bacon, Descartes, Fabricius, and Harvey – are university-based philosophers. Indeed, one might easily come to think there *was* no university tradition of philosophy from reading Gassendi's later work.

A common target in the *Syntagma* is the neo-Platonism of advocates of the world soul. The most developed argument against the world soul occurs in the initial sections of the *Physics*, where Gassendi is setting up his claim that atomic motion is the sole created cause by arguing against other extant accounts of created activity.[14] The *anima mundi*, as Gassendi understands it, explains created activity by informing or ensouling particular things: It is "as it were scattered and divided into particles which result in particular souls or forms, not only of men but also of beasts and even of plants, metals, stones and everything in general" (1.334a). However, Gassendi says, "this is not only an impious thing to say but also a most absurd one: as if indeed an incorporeal, immense, and nowhere-not-existing Entity could be separated, carried along and caught up by a body!" (1.334a). The impiety, as will become clear in the argument against Robert Fludd, lies in the fact that those who accept a world soul must hold that all souls, whether of men or of rocks, have the same status vis-à-vis eternity. The absurdity is supposed to lie in the claim that the world soul, an incorporeal being that exists everywhere, can be a genuine principle of activity in individual bodies. For this would require the world soul to have parts that move around with particular bodies. However, the advocate of the world soul must presumably think of it as genuinely immanent in bodies, else it does not solve the problem of secondary causation but merely makes the impious claim that there is a transcendent principle of causation between God and bodies. If the world soul is not genuinely in bodies, then bodies themselves cannot be secondary causes. Moreover, Gassendi adds, if the world soul is incorporeal, "it cannot be grasped how . . . it can be applied to bodies so that it impresses impulses on them" (1.334b).

Gassendi is careful to note that the objection is to a world soul that is understood as an incorporeal being. He is not concerned to argue against

[14] In this and the preceding section, I consider the two most important such accounts, hylemorphism and the world soul. However, Gassendi also objects to a number of views on which matter itself is active in virtue of being qualified in different ways – by the elemental qualities hot, cold, wet, and dry, for instance, or by the chemical qualities salt, sulfur, mercury, earth, and water. For a fuller account of these objections, see LoLordo, "The Activity of Matter in Gassendi's Physics."

those who, by holding that the world soul is something like vital heat, an affection of bodies, hold that it is corporeal. Such authors really agree that the efficient principle in the created world is material.

Gassendi's objections to the doctrine of the world soul are presented most explicitly in his *Examen Philosophiae Roberti Fluddi Medici* (*Examination of the Philosophy of the doctor Robert Fludd*), so let us turn to that work. Who *was* Robert Fludd? As well as being a physician with interests in judicial astrology and iatrochemistry, he was interested in the Cabala and the theory of chemistry, and he argued that there was an *anima mundi* and sketched out a Paracelsan reading of Genesis.[15] Fludd's system is in the tradition of Ficino and Patrizi, like them using elements from neo-Platonism, the Chaldean oracles, and the *Corpus Hermeticum.* It is Fludd's account of divine creation that is the source of his engagement with Gassendi, an engagement mediated by Mersenne, whose *Quaestiones in Genesim* is an attack on Fludd's 1617 *De Macrocosmi Historia.*[16] Mersenne was incensed by Fludd, calling him an evil magician and a heretic. This was part of a larger attack on various forms of occultism supposed to subvert religion. Mersenne's attack encompassed targets such as Campanella, Paracelsus, Pomponazzi, Bruno, and Vanini, who like Bruno had been executed for atheism – in Vanini's case only four years previously.

Fludd defended himself against Mersenne's charges in 1629's *Sophia cum moria certamen* (*The Combat of Wisdom with Idiocy*). Mersenne asked Gassendi to write a response on his behalf, and Gassendi did so in the critical but relatively restrained *Examen.* As an instance of this relative restraint, Gassendi says that Mersenne should not have called Fludd a witch, heretic, magician, or atheist – most clearly not the latter, since he is constantly talking about the divine nature and quoting the Bible. Moreover, although the views expressed are dangerously heterodox, it is not clear that Fludd realizes this (3.240b–1a). Fludd is making a mistake that needs correction, Gassendi suggests, rather than proposing heretical views.

[15] For accounts of Fludd's life, works, and significance, see Huffman, *Robert Fludd and the End of the Renaissance,* or Fludd, *Robert Fludd and His Philosophical Key* (editor's introduction). For the quarrel between Fludd and Gassendi, see also Brundell, *Pierre Gassendi,* 110–12, or Clericuzio, *Elements, Principles and Corpuscles,* 71–4.

[16] Mersenne also devotes some attention to Fludd's *Tractatus Apologeticus Integritatem Societatis de Rosea Cruce Defendens,* also of 1617. Two sequels to *De Macrocosmi Historia* appeared in 1618 and 1619 and the three parts were printed together as *Utriusque cosmi maioris scilicet et minoris metaphysica, physica atque technica historia.*

The charmingly titled *Doctor Fludds answer unto M. Foster, or, The squeesing of parson Fosters sponge* contains Fludd's replies to both Mersenne and Gassendi.[17] He answers the two men more or less in kind, contrasting Mersenne's work with his "honest dealing and morall" (16) friend: Mersenne is a "roaring, bragging, and fresh-water Pseudophilosopher, while Gassendi is "an honest and well conditioned Gentleman . . . not passing beyond the bounds of Christian modestie, but striking home with his Philosophicall argument" (18–19). The comparison is chiefly intended to discredit Mersenne. Nevertheless, it is interesting to read this reception of Gassendi's critique and contrast it with the tone of Descartes's *Replies*. For Gassendi's judgment of Fludd is certainly no less harsh than his judgment of Descartes.

Gassendi, like Mersenne, objects to Fludd's reading of Genesis because it elevates chemistry to the status of religion (3.259b), an abuse of Holy Scripture that is common to Fludd "and other Alchemical Physicians" (3.231b). Gassendi is not arguing against chemistry in general here: Indeed, he takes pains to point out the respect he has for it, granting, for instance, that we should be able to turn lead into gold although no one has yet managed to do so (3.259a). More generally, chemistry is useful in medicine and in the investigation of nature, where "even if we cannot penetrate all the way into the inmost depths of natural things," if something about their "essences, discriminations, powers, actions, modes of acting . . . and contexture" can be known, it can be known through chemistry (3.259a). Gassendi objects not to chemistry in general, then, but to Fludd's presumption that it can form the basis of a complete metaphysics and theology.[18]

Gassendi also objects to the tendency toward secrecy he sees in Fludd, a tendency that he compares with Mersenne's alleged openness and concern with sensible things (3.213). This tendency toward secrecy is one of the hallmarks of the obscure philosophy, along with another feature of Fludd's writing, a reliance on symbolism. Discussing Fludd's pyramids of relative quantities of form and matter in the different spheres, for instance, Gassendi writes that "it is obvious . . . that one should not look

[17] In *Hoplocrisma-spongus, or, A sponge to wipe away the weapon-salve*, Foster had argued that the efficacy of the weapon-salve is due to its diabolical nature and had attacked Fludd for integrating such evil magic into his system. Fludd writes that Foster "talks with Mersennus the Fryer his tongue, and therefore is but Mersennus his Parrot" (Fludd, *Doctor Fludds answer unto m. Foster, or, The Squeesing of parson Fosters sponge*, 7).

[18] Monnoyeur, "Matter: Descartes versus Gassendi," helped draw my attention to the importance of Gassendi's positive views on chemistry.

for pyramids of this sort *in rerum natura*, since they are only symbols" (3.228a). Equally worrying are Fludd's hostility to observational astronomy, whose legitimacy he allows only when it contributes to medicine, and his critiques of Gilbert and Copernicus (3.223a). Thus, Gassendi's choice of an appendix containing some brief astronomical notes for Mersenne is pointed.

My central concern in discussing the *Examen Fluddi* is the doctrine of the world soul. Gassendi tells us that he rejects the theory of the world soul in all its variants, and I take it that his arguments against Fludd's world soul are supposed to be generalizable. But it is easiest to begin from the role and function of the world soul within Fludd's particular metaphysic. This metaphysic is developed through a system of triads. One triad is matter, form, and spirit, where spirit is neither material nor immaterial but instead acts as the link between the two. This triad is associated with the Trinity: matter is God the Father, form Christ, and spirit the Holy Spirit. The Trinity in turn is associated with the three realms of the macrocosm and the corresponding three realms of the microcosm: the *Empyreum* or high heavens, corresponding to the human head; the *Aethereum* or aerial realm, corresponding to the thorax; and the *Elementare* or earth, corresponding to the belly (3.226a).

The three parts of the microcosm are ruled by different faculties and correspond to different ontological principles, the head being ruled by the *anima* or intellectual faculty and thus being associated with form, the thorax with the *medium uniens* or spirit and with the vital faculty, and the belly with matter and the sensitive faculty (3.228a). Just as the human being, the microcosm, is governed by a soul, so the world as a whole is governed and animated by an *anima mundi*, "the union of immortal rays of light emitted from eternity with the most subtle and meritorious portion of humid nature or water," so that the world soul is a unity derived from an aggregate (3.221a). Just as the soul informs the whole matter of the human body via the spirit, the world soul informs the whole matter of the world (3.221b). Fludd identifies the world soul with the great angel Mitatron of the Cabala, the sun (which is its seat), the Incarnate Word, the Messiah, Christ, the Savior, and the Philosopher's Stone (3.221b). This last identification is especially worrying to Gassendi, who points out that since Fludd means by the Philosopher's Stone just "that special gold or golden substance that would give eternal youth and immortal life," he is ascribing the power of salvation to this special gold, "to the damage of sacred Religion" (3.222a).

However, Gassendi's chief source of worry is the equation of the world soul with God – and hence, given the parallelism between macrocosm and microcosm, the equation of the *human* soul with God. First, he argues that this makes God entirely immanent, thus implying a number of theologically unacceptable attributes. Second, intolerably, it puts God and the human soul on the same level. Since the world soul is a composite, God himself must be composite, a conclusion that is incompatible with the divine simplicity. It implies that God must be made out of parts, which would be an imperfection, and that some of his parts are agglutinations of matter, which contradicts God's immutability. It implies that the different parts of God will have different locations, which is unacceptable because God is supposed to wholly exist at all places at once. And it makes God subject to rarefaction and contraction (3.236a–7a). (The similarities to Bayle's and Leibniz's objections to Spinoza are striking.) Fludd's theory of the world soul in effect construes God as in *every* part of matter because every part is informed. Hence, Gassendi says, if Fludd were right, then there would be nothing wrong with idolatry because the essential form of a stone or a log would be God (3.237a).

Moreover, Gassendi objects, the strict correspondence between microcosm and macrocosm forces upon Fludd the view that the human soul has existed from eternity just as the world soul has. However, this view is inconsistent with the "common opinion of the Theologians" that the soul is infused into the body at the moment of its creation (3.226b). Indeed, Fludd's doctrine of the world soul, as Gassendi understands it, implies that *no* soul can be any more or less mortal than any other (3.237a). And it implies further that there is no action of creatures, only the action of God – a consequence that is unacceptable because, by eliminating secondary causation, it makes God the cause of virtues and vices alike (3.237b).

Of course, in attacking the world soul, Gassendi is not simply attacking Fludd but has bigger targets in mind as well: "Pythagoras, according to Stobaeus, and Plutarch . . . and also Plato, and in a word whoever, asserting that there is an *anima mundi*, judges that all special forms of things are little parts of the world soul, from which parts all efficacy in things comes" (1.333a–b). One influential late Renaissance *anima mundi* theory is found in Marsilio Ficino's 1474 *Platonic Theology*.[19] Ficino explains that one of

[19] Although I am not aware of any proof that Gassendi read this work, he was certainly familiar with Ficino. As we saw in Chapter 1, he owned some of Ficino's translations of Plato when he died.

his goals is to show that "besides this inert [*pigram*] mass of bodies which the Democriteans, Cyrenaics and Epicureans limit their consideration to, there exists some efficacious quality or power."[20] Thus, in responding to a view like Ficino's, Gassendi is intervening in a preexisting debate between neo-Platonists and Epicureans.

In the chapter "Body does not act of its own nature," Ficino explains that body – by which he understands matter and its extension, quantity – cannot act because "it is characteristic of matter only to be extended and to be affected, and extension and affection are passions."[21] Thus, "all power of acting must be attributed to an incorporeal nature,"[22] and "if bodies appear to act in any way, they do not do so by virtue of their own mass, as the Democriteans, Cyrenaics and Epicureans supposed, but through some force and quality implanted in them."[23] Therefore, Ficino argues, "there must be a certain incorporeal substance [or form] present in and ruling over all objects,"[24] and "the powers and activities of qualities are based on the power and activity of such an incorporeal form."[25] This incorporeal form in the end "is nothing other than soul," as "soul is both movable of itself and bestows on bodies the imprint of its movement."[26] Ficino assigns souls to individual humans and animals and the twelve heavenly spheres, but at the highest level these are subsumed in "the single world soul," which is thus the ultimate source of all apparent activity in bodies.[27]

Ficino's account of the world soul differs from Fludd's in that he does not suggest that the *anima mundi* is God. But although Gassendi's objections are directed against a world soul that is identified with God, many apply equally well to any view on which there is an incorporeal soul divided into parts that serve as the souls or forms of particular things:

(a) It implies that all souls – souls of men and souls of lead – are equally mortal or equally immortal;
(b) It requires the division of an incorporeal thing, which Gassendi cannot accept since he holds that the only sort of real (as opposed to merely conceptual) division is division by cutting and separating, processes that require resistance or impenetrability;

20 Ficino, *Platonic Theology*, 15 (Section 1.1).
21 Ibid., 19 (Section 1.2).
22 Ibid., 21 (Section 1.2).
23 Ibid., 23 (Section 1.2).
24 Ibid., 37 (Section 1.3).
25 Ibid., 41 (Section 1.3).
26 Ibid., 281 (Section 4.1).
27 Ibid., 295 (Section 4.1).

(c) It requires the world soul to have parts that move around with different bodies, which is incompatible with its being one entity and thus shows that there is a contradiction in the theory;
(d) If the world soul is incorporeal then "it cannot be grasped how . . . it can be applied to bodies so that it impresses impulses on them, since it cannot contact them, lacking that tangibility [*tactu*] or bulk [*mole*] by which it could touch them" (1.334b).

Problems (a) through (c) apply whether or not the world soul is identified with God. The reader may at first think that objection (d) simply begs the question. For, she might reason, surely no one who believes that an *anima mundi* is the locus of activity in nature would accept the principle that all action is the impression of impulse by contact. Indeed, Gassendi's main argument for this principle is simply the overarching argument that there is no other acceptable way to make sense of secondary causation – an argument that proceeds by rebutting forms, the *anima mundi*, and various qualities beyond size, shape, and motion as sources of activity.

However, one need not read objection (d) as relying on the assumption that *all* action proceeds through the impression of impulse. All Gassendi needs is the much weaker claim that *some* physical causation is the impression of impulse through contact. This weaker claim is uncontroversial. The objection then is this. A world soul that is incorporeal could not impress impulse on anything. This is incompatible with the world soul's being the only principle of activity within nature. But it is essential to the doctrine of the world soul that *all* creaturely activity is really the activity of the world soul.

Objection (d) might bring up a second worry in the reader's mind as well. If Gassendi claims that the incorporeal cannot affect the corporeal, doesn't he commit himself either to the entirely unacceptable view that God is corporeal or to the equally unacceptable view that God cannot affect the created world? I described (d) as an objection to the world soul only when construed as distinct from God – but, the reader may ask, doesn't it apply equally to God himself?

However, Gassendi is careful to note that his claim that the incorporeal is unable to affect the corporeal is limited to created things and does not apply to God, for three reasons.[28] First, God, unlike created things,

[28] In this section of the *Syntagma* – although *not* in later ones – Gassendi takes this doctrine to apply to the incorporeal human soul, which elicits only "intellectual or mental and incorporeal" acts, rather than having causal influence on the body. Now, this is a standard scholastic view, but not one that Gassendi can successfully assimilate. For it relies on the rational soul affecting the body indirectly, via the mediation of the sensitive soul. At

has unlimited power or *virtus*. Second, God, unlike other incorporeal things, is present everywhere.[29] Finally, unlike the action of corporeal things, divine action is *nutu*, by command, rather than *motu*, by motion. Gassendi is committed, here as throughout his work, to an account of God's activity on which divine activity differs radically from creaturely activity and hence cannot be fully understood by analogy with creaturely activity.

It is interesting to see that Gassendi brings up very similar worries against Cartesian physics. He argues that the Cartesian thesis that the essence of body is extension would require reintroducing an incorporeal source of bodily activity in order to make sense of secondary causation:

> Concerning body, I note only this, that if all its nature consists in its being an extended thing, then all action and all faculty of acting will be beyond corporeal nature because extension is merely passive and he who says that a thing is merely extended says among other things that it is not active. Hence there will be no action, no faculty of acting in bodies: and from where will [the action or faculty of acting] come from? From an incorporeal principle? But that which is incorporeal is only thinking and cannot elicit corporeal actions. Or from a corporeal principle? But that which is corporeal is only extended and not really something that acts... And you will have to go back to forms or *differentia* or whatever else you want. (3.305b; compare 3.284b)

That is, Gassendi argues, if to be corporeal is just to be extended then nothing corporeal can have activity. And because he always operates on the assumption that there *is* genuine activity in the created world, it follows that Descartes will need an incorporeal source of activity within physics. Thus, the Cartesian will be forced to resuscitate the unacceptable forms of the Aristotelians or the equally unacceptable notion of a world soul that activates matter.[30]

points Gassendi suggests that the corporeal soul, which we humans share with plants and animals, serves as a *nexus* between body and incorporeal soul. However, this suggestion is in the end no real help because the corporeal soul is itself entirely corporeal, though composed of particularly fine and "spiritual" corpuscles – and hence the problem about the incorporeal being able to affect the corporeal persists.

[29] I discuss this claim, which Gassendi sometimes intends entirely literally, further in Chapter 10.

[30] Gassendi does not add the possibility that the Cartesian could be forced into occasionalism instead, although it is easy to add this with hindsight. Rather, he assumes that the Cartesian will ultimately have recourse to some immaterial principle in order to avoid occasionalism.

The argument against the world soul has two main roles in the programmatic sections of the *Syntagma*. First, by attacking a central explanatory concept of Renaissance neo-Platonism, it helps motivate and clear the way for Gassendi's atomism. Second, more specifically, by ruling out a major alternative view of created causality, it paves the way for Gassendi's introduction of atomic motion as the only genuine locus of causality. In this, it is paired with the argument against Aristotelian forms as causes discussed in the previous section.

These two arguments are accompanied by various objections to taking either the primary or elemental qualities or the various secondary qualities as sources of causality.[31] The view that the qualities of matter are the efficient principle is really the view that matter, as qualified, is the efficient principle, and hence Gassendi's arguments here can also be read as directed against *minima naturalia* theories on which there are ultimate "atoms" of a limited number of different kinds. For he makes essentially the same point against all theories on which matter is active in virtue of being qualified in specified ways, whether the qualities are heat, cold, wetness, or dryness; salt, sulfur, and mercury; or primeval heat and light. In the end, Gassendi thinks, it is fine to call any of these things active principles, so long as we allow, first, that they are wholly corporeal and, second, that they are explicable in terms of corpuscles or molecules (small masses of atoms).

Consider a chemical example. Gassendi writes that the five-element view, on which everything is composed of salt, sulfur, mercury, earth, and water, is plausible. For a generally accepted chemical principle holds that "every thing consists in those things into which it is resolvable," and chemists have shown that some things, for example certain metals, are indeed resolvable into the five elements (1.245a). However, he thinks that these five elements could themselves be further resolved, ultimately into atoms: "I remain silent here about what could be added concerning the resolution of those five principles into their seeds and finally into atoms, for the matter should be understood from the things that will be said later" (1.245b). "Resolution" is, of course, a chemical term, but Gassendi's assumption that whatever a body can be resolved into are its true components may also rely on the scholastic axiom *in quae dissolvi possunt*

[31] In Gassendi's usage, the primary or elementary qualities are the four Aristotelian elemental qualities (hot, cold, wet, and dry), while any further qualities are called secondary. This usage is entirely traditional. See, for example, Goclenius's *Lexicon philosophicum*, article "Qualitas."

composita, ex iisdem coaluerunt (that which composites can be dissolved into is that which those composites are made from).[32] Gassendi's suggestion is that, at least in theory, the same sorts of processes the chemists have used to arrive at the elements could be used to decompose the chemical elements into atoms.

The claim that the chemical elements are further resolvable is hardly original. Mersenne, for instance, suggests in *La vérité des sciences* that the three elements salt, sulfur, and mercury could be further resolved into earth and water, van Helmont's basic constituents.[33] It is the claim that the elements are resolvable into atoms that is interesting, and Gassendi makes this claim about a number of different, allegedly primitive constituents of nature. He recognizes, for instance, theoretically subdivisible corpuscles of salt, gravity, magnetism, niter, light, and so on, all of which are composed of atoms. Now if this is what the elements are, then there is little reason to minimize the number of elements. There would not even be any gain in simplicity. Gassendi's appeal to a variety of different types of corpuscles is thus consistent with an account of individual atoms as differing only in terms of size, shape, and motion.[34] He applies the strategy of reinterpreting elements as composites of atoms to a whole range of previous matter theories, and if the strategy works, his atomism can capture the successes of almost any previous theory.

Descartes and the *Meditations*

Scholastic Aristotelianism and the doctrine of the world soul were central ingredients of the intellectual climate in which Gassendi was formed. The same cannot be said of Cartesianism: Gassendi was four years older than Descartes, and by the time he wrote the *Objections* to Descartes's *Meditations* he had already arrived at most of his fundamental views. We know that Gassendi read the *Discourse, Optics, Meteorology*, and *Geometry* and that he was impressed by the mathematical results therein, though he does not mention what he thought of the *Discourse* (3.275b). He later recognized

[32] Newman, "Experimental Corpuscular Theory," 324.

[33] Mersenne, however, differs about what the chemical elements could be resolved into: "As for the alchemists' principles, I hold that salt, sulfur and mercury are not first principles because they can be resolved again into two bodies, namely into earth and water." (*La vérité des sciences*, 56).

[34] I think it is also intended to be consistent with the claim that all change results from local motion. However, it is clear by the end of the *Syntagma* that this claim can only with difficulty be applied to living beings.

the new prominence of Descartes, who is, for instance, the only contemporary figure discussed at length in Gassendi's history of logic and in his account of the material principle of nature.[35] However, Cartesianism served not as an influence but as a foil against which Gassendi could sharpen his own views: The main issues of the *Fifth Objections* are central for the theory of cognition developed in the *Syntagma*.

Gassendi came to write a set of *Objections* when Mersenne chose him as one of the people among whom to circulate the *Meditations*. Descartes was then familiar with – and apparently admired[36] – Gassendi's astronomical work, although as far as I know they had never met. In his *Meteorology*, Descartes had used what were apparently Gassendi's observations from *Parhelia sive soles* without giving him credit, and Gassendi had objected to this.[37] It is unclear whether any lasting bad feeling arose as a result of this, or whether the hostile tone of the *Objections* and *Replies* derives from philosophical disagreement alone.

The *Fifth Objections* were composed in April and May of 1641. Gassendi begins by noting that his worries "did not concern the actual results [Descartes had] undertaken to prove, but merely the method and validity of the proof" (3.273b).[38] He must have had in mind the advertised conclusions of the *Meditations*, "the existence of almighty God and the immortality of the soul." For he also disagreed with Descartes's conclusions in such theologically neutral cases as the nature of body and the role of the pure intellect.

Because Gassendi's first confrontation with Cartesian philosophy was with the *Meditations* and not the *Principles*, the concerns he expresses about Cartesianism are at bottom concerns about Descartes's methodology and the theory of cognition he takes it to rely on. These concerns are embodied in his worries about the notion of clear and distinct perception and the innatism he takes it to rely on. Looking at how Gassendi objects to the Cartesian account of cognition and knowledge

35 However, Gassendi apparently planned to write a chapter on Hobbes's logic as well. See the letter of Abraham du Prat to Hobbes asking him, on behalf of the person in charge of the Gassendi edition, to write a summary of his logic to insert "at the place where he had intended (if death had not prevented it)." Hobbes, *The Correspondence*, 1.343.

36 See the letter to Mersenne of January 1630 (Descartes, *Oeuvres de Descartes*, 10.18).

37 Rodis-Lewis, *Descartes*, 84.

38 I have benefited greatly from consulting the translation of the *Fifth Objections* by John Cottingham, Robert Stoothoff, and Dugald Murdoch (in Descartes, *The Philosophical Writings of Descartes*) and the partial translation of the *Counter-Objections* by Craig G. Brush (in Gassendi, *The Selected Works of Pierre Gassendi*).

will also serve as a useful introduction to Gassendi's own philosophical methodology and his positive account of cognition.

Gassendi understands the criterion of clarity and distinctness as supposed to fulfill a dual role – more or less the same role, we shall see, that he takes the evidence of the senses to succeed in filling. The clarity and distinctness of a perception is something a perceiver is supposed to have immediate, incorrigible introspective access to, and all clear and distinct perceptions are supposed to be true. However, Gassendi objects that Descartes has failed to establish that all perceptions introspected as clear and distinct are actually true. In fact, he argues, we know that some allegedly clear and distinct perceptions must be false. For one thing, we recognize cases when we apparently perceived one thing clearly and distinctly and later perceived another incompatible thing clearly and distinctly. Gassendi gives the example of asymptotes. As a child he thought that two lines continually approaching each other more closely must meet, but he later learned this was false, and still later gave up having any opinion whatsoever as a result of considering the nature of mathematical propositions (3.314b). For another, different people have inconsistent, allegedly clear and distinct perceptions (3.315a). This is supposed to show that anyone who holds that all clear and distinct perceptions are true must also hold that there can be mistakes about whether a perception is clear and distinct, and hence that we need a mark by which *genuinely* clear and distinct perceptions can be picked out:[39] "What you should be working on is not so much confirming this rule which makes it so easy for us to take the false for the true, but instead proposing a method to guide us and teach us when we are mistaken and when not, in the cases where we think we clearly and distinctly perceive something" (3.315a). Since Descartes has not provided any such mark, Gassendi argues, his method is all but useless. Gassendi is not merely bringing up the notorious problem of the criterion here. Rather, he is arguing that there is particular need for a criterion in this case. This contrasts with the case of sense perception: On Gassendi's account of sense perception, there can be no inconsistency or falsity in the reports of the senses, and hence there is no need for a criterion of truth for the senses.

[39] Descartes responds that "it can never be proved that [such people] clearly and distinctly perceive what they so stubbornly affirm" (*Oeuvres de Descartes*, 7.361). Gassendi simply does not take this response seriously, writing that "the fact that these men go to meet death for the sake of some opinion appears to be a perspicuous argument that they perceive it clearly and distinctly" (3.317a).

In response, Descartes points out that clarity and distinctness are supposed to be features only of the perceptions of people who have freed their minds from the senses by properly following the meditative method. Those people cannot be mistaken about what is clearly and distinctly perceived – but since Gassendi is still "mired in the senses," inconsistencies among his allegedly clear and distinct perceptions are irrelevant.[40] Because those who have abandoned preconceived opinions and freed themselves from the senses will never disagree, clarity and distinctness needs no mark beyond the compulsion it engenders. The method for determining which perceptions are genuinely clear and distinct has already been given.[41] This response initially seems unhelpful. Gassendi asked for a method for distinguishing genuinely clear and distinct perceptions from belief-compelling, but not clear and distinct, acts of the mind. Descartes responds that a method has been given, as follows: Eliminate all preconceived opinions, list the principal ideas, and distinguish the clear ones from the obscure ones. But how can we carry out this last step? Isn't distinguishing clear principal ideas from obscure ones just a special case of distinguishing the genuinely clear and distinct? Descartes's answer, I take it, is yes – but it is a case where error is impossible. For error in seeing what's perceived clearly and distinctly results from the contagious effect of obscure and confused ideas acquired before the age of reason.[42]

Because Descartes's response relies on the claim that the method of meditating eliminates conflicts within what is apparently clear and distinct, the argument now shifts to the possibility of doubt itself. Gassendi construes the First Meditation's doubts as a tool intended to allow us to isolate the clear and distinct perceptions of the intellect. Thus, he understands doubt as centrally important (3.279b). On his reading, the call for doubt is an experiment intended to convince the meditator of the distinction between intellectual or innate ideas and sensory cognition. Gassendi summarizes – or, perhaps better, parodies – Descartes's argument for innate ideas of the intellect as follows:

> He who, having previously known things, considers himself in a state of ignorance as a result of his pursuit of the divestment of ideas, readily recognizes innate ideas, i.e. those which we would have in a state of ignorance. . . .
>
> Now I, having previously known things, consider myself in a state of ignorance as a result of my pursuit of the divestment of ideas.

[40] Descartes, *Oeuvres de Descartes*, 7.379.
[41] Ibid., 7.361–2.
[42] Ibid., 7.164.

Therefore I readily recognize innate ideas, i.e. those which I would have in a state of ignorance. (3.320a)

Gassendi interprets Descartes as claiming that the fact that a certain class of ideas – the ideas of body, extension, shape, quantity, and the like[43] – remains after all the ideas acquired through the senses have been thrown off shows that those ideas are innate and pertain to an independent faculty of the intellect. If this is right, then Descartes's argument for purely intellectual ideas relies on our actually having carried out the divestment of all sensory ideas. However, Gassendi argues, it is impossible for us to rid ourselves of all the ideas acquired through the senses: indeed, impossible even to rid ourselves of sensory *beliefs*.[44] Picking up the ancient antiskeptical tactic of arguing that the skeptic cannot live his doubts, Gassendi argues that the method of doubt is "merely verbal" and that merely verbal doubt is unhelpful: "If [you hold preconceived opinions to be false] in name and as a jest, then this does not make you advance at all toward the correct perception of things. For whether you do this or not, the mind is not affected at all, and it is not helped either to perceive or to perceive more clearly" (3.280b). Merely verbal doubt cannot make the mind capable of clear and distinct perception because it cannot erase the sensory beliefs that Descartes thinks distort judgment. Without genuinely doubting the beliefs of the senses, we cannot tell whether certain intellectual beliefs rely on them without our knowing it.

But why is doubt about beliefs like "I am sitting in front of a fire" supposed to be impossible? Gassendi gives us three reasons. His first reason is experiential: Despite a great deal of effort, he himself has not succeeded in freeing his mind from preconceived opinions by meditating along with Descartes. The method, Gassendi says, didn't help him any. It is easy to dismiss this flat-footed response, but if we think of Descartes's project as laying down steps to follow for the augmentation of learning, it is a serious failing if dutiful emulators cannot manage to follow these steps.

Second, Gassendi argues, the way memory operates makes abandoning all beliefs impossible. Even if we want and have reason to abandon

[43] Ibid., 7.20 and 7.28.

[44] The two philosophers distinguish ideas from beliefs or judgments in different ways. For Descartes, an idea can be either termlike or propositional, and judgment is a matter of giving or denying assent to the content presented in an idea. For Gassendi, ideas are termlike, and judgments are the joining of two ideas by *is* or *is not*. This means they are talking past each other at certain points.

the beliefs of the senses entirely, the influence of old beliefs remains in memory: "[B]ecause the memory is like a storehouse of judgements that we have previously made and deposited in its keeping, we cannot cut it off at will ... judgements already made persist so strongly by habit, and imprints similar to those of a signet are so fixed, that it is not in our power to avoid them or erase them at will" (3.279a). Here, Descartes might agree that in some sense we still "retain the same notions in memory" when we withhold assent.[45] But because he thinks we have the capacity to form judgments independently of the traces held in the corporeal memory, the existence of such traces does not prevent us from withholding assent from certain judgments voluntarily, but only makes it somewhat more difficult for us to do so.

Gassendi's third reason for thinking that Cartesian doubt is psychologically impossible is simply that there are some types of belief that we cannot help assenting to given the evidence available to us:

> [T]here are some preconceived opinions that can be changed and there are likewise some that cannot ... because every judgement refers to some object as it appears to the mind, if the object always appears the same way – as the sun always appears round and shining, or as the meeting of two straight lines always appears to form two right angles or two angles equal to two right angles – then we always make the same judgement ... [these] judgements cannot be changed (because what appears in the object is not changed), and therefore ... the mind cannot escape them. (3.279b)

Ideas, the claim is, determine belief in such a way that if the appearance of something to the mind is always the same, we cannot change our beliefs. Gassendi thinks this accords with introspection. However, the argument also relies on the substantive theses about cognition to which the next two chapters will be devoted.

45 Descartes, *Oeuvres de Descartes*, 9.204.

3

Skepticism, Perception, and the Truth of the Appearances

Skepticism is both a tool and a threat for Gassendi. Richard Popkin's characterization of Gassendi and Mersenne as mitigated or constructive skeptics – thinkers who reject Aristotelian *scientia* and its claims to know essences on the one hand and the skeptical claim that nothing can be known on the other – is well known.[1] It captures an important aspect of Gassendi's thinking, especially in his early works. However, in this as in so many other matters, there is a development in Gassendi's thinking. The early *Exercitationes*, Popkin's chief text, emphasizes skepticism as a weapon against Aristotelianism. In contrast, the system of canonic provided in the *Syntagma* outlines a positive program for achieving knowledge, one that responds to Gassendi's earlier skepticism.

In *La vérité des sciences*, Mersenne uses Pyrrhonist arguments to undercut chemists and other opponents of "the Christian Philosopher," while rejecting Pyrrhonism about knowledge in general. His chief strategy against an overarching Pyrrhonism is to place strict limits on the scope of the certain – allowing only mathematics and the articles of faith to count – and to construe the majority of natural philosophy as merely probable. Similarly, Gassendi uses Pyrrhonian techniques to rebut Aristotelian metaphysics and physics. At the same time, he carefully limits the scope within which Pyrrhonism can legitimately be deployed. This tactic is evident as early as the *Exercitationes.* We have already seen Gassendi's attack on knowledge of essences and *scientia,* but even in the *Exercitationes* Gassendi allows that we have some knowledge, although it does not

[1] Popkin, *The History of Scepticism,* 129 ff. For some important differences between Mersenne's and Gassendi's skepticism, see James, "Certain and Less Certain Knowledge."

count as *scientia*: Many of the doctrines of the particular arts are known, and so are the appearances.

Gassendi's later work is rather less skeptical. By the *Syntagma*, he grants the possibility of knowledge of various natural philosophical claims. We know, for instance, that atoms exist and move through the void, and we know that bodies are congeries of atoms, variously disposed. Since the *Syntagma* makes knowledge-claims, one of its primary tasks is to explain how such knowledge can be acquired. This task is begun in the *Logic* and completed in the chapters of the *Physics* that treat the human cognitive capacities.

Gassendi transforms the Pyrrhonian opposition between essence and appearance into a distinction between a hidden inner structure and the manifest sensible qualities that flow from it. His account of cognition is central to this transformation. For Gassendi eliminates conflict among appearances by showing how the disposition of our sensory organs and cognitive system contribute to the content of perception. This tactic is Epicurean in inspiration if not in detail, for the denial of genuine perceptual conflict, error and disagreement was a cornerstone of Epicureanism and its claim that all perceptual appearances are true.[2]

One strand in Gassendi's rejection of Pyrrhonism, then, is an attempt to cut down on the amount of perceptual conflict and disagreement we experience. Conflict among the appearances is crucial to Pyrrhonism, as Sextus makes clear: "Skepticism is an ability to set out oppositions among things which appear and are thought of in any way at all, an ability by which, because of the equipollence in the opposed objects and accounts, we come first to suspension of judgment [*epoche*] and afterwards to tranquillity [*ataraxia*]."[3] Judgment is suspended when conflicting evidence is set out on both sides and the opposing sides have "equipollence," or "equality with regard to being convincing or unconvincing," a state that takes place when "none of the conflicting accounts takes precedence over any other as more convincing."[4] Pyrrhonian doubt requires not merely the existence of some evidence on each side but the felt *equality* of evidence.

Pyrrhonians offered a number of techniques for coming up with equipollent arguments. However, the techniques most important to

[2] See for example Diogenes, *Lives of Eminent Philosophers*, 10.50.
[3] Sextus, *Outlines of Scepticism*, 8 (Section 1.4).
[4] Ibid., 10 (Section 1.4).

Gassendi involve the Ten Modes – tropes that centrally involve perceptual relativity and disagreement. Such perceptual relativity can be due to physical differences between individuals or groups, as when "Indians enjoy different things from us";[5] differences between the senses, as when "paintings seem to sight to have recesses and projections, but not to touch";[6] a difference in position or place, as when "the same tower appears from a distance round, but from close at hand square";[7] and so on for the other seven modes. Thus, if Gassendi can eliminate or reduce the degree of opposition between different perceptions, he can defeat what he takes to be the most pressing argument for skepticism about the senses.

Another strand in Gassendi's rejection of Pyrrhonism is his attempt to provide a logic of probability. This is the task of his theory of signs, which regulates the practice of making inferences from the manifest qualities of things to the underlying essences of the things bearing those qualities. Because essences are not sensed, we cannot know them in the same way we know appearances; thus, given Gassendi's theory of cognition, inference is the only path available. We gain probable knowledge of essences by understanding manifest qualities as signs of the underlying corpuscularian structure that causes them – and in cases where only one structure could give rise to those qualities, we can even gain certainty by using signs. A prominent example is the claim that the world consists of atoms in the void, which is supposed to be the only way the motion evident to perception could be possible. I discuss the elimination of perceptual conflict here, and the theory of signs in the following chapter.

Gassendi and the Pyrrhonians

Throughout his career, Gassendi was impressed by the tropes that lead to skepticism about the senses as well as those supposed to produce skepticism about the power of reason. However, he argues that concerns about the reliability of the senses can be rejected if we correctly understand the Epicurean claim that all perceptions are true. Maintaining this claim requires eliminating even the possibility of conflict between different perceptions by reinterpreting the content of perception. Hence the

[5] Ibid., 80 (Section 1.14).
[6] Ibid., 92 (Section 1.14).
[7] Ibid., 118 (Section 1.14).

plausibility of Gassendi's claim that the appearances are always true depends in large part on the success of his account of perception.

Gassendi's knowledge of Pyrrhonian skepticism, like his knowledge of ancient Epicureanism, came from a number of sources: Diogenes Laertius, Cicero's *Academica*, Augustine's *Contra Academicos*, and Sextus Empiricus's *Outlines of Scepticism*. Gassendi's use of these texts was not new. Sextus's *Outlines of Scepticism* – first published in Latin translation by Henri Étienne in 1569 – was crucial for the sixteenth-century revival of skepticism. The devotion of the young Gassendi to some of the chief sixteenth-century new Pyrrhonians is evident. In a letter of 1621, he lists Pierre Charron and Michel de Montaigne as among his favorite writers (6.2a). The preface to the *Exercitationes* adds Juan Luis Vives, Petrus Ramus, and Pico della Mirandola. Indeed, Gian Francesco Pico della Mirandola's use of arguments from Sextus in his *Examen vanitatis* may have provided Gassendi with a model for his own skeptical attack on Aristotelianism in the *Exercitationes*.[8]

Gassendi begins the *Exercitationes* by endorsing "the *akatalepsia* of the Academics and Pyrrhonists" (3.99, no columns). This is an unusual remark: He typically distinguishes the Academics, who say that "truth cannot be found" and "judge that all things are incomprehensible," from the genuine skeptics who persist in inquiry. He took the distinction to be important as early as 1634, where it is clear that he favors Pyrrhonism.[9] While the Academics take *akatalepsia* as the end-point of inquiry, the Pyrrhonians, on Gassendi's account, "neither state that truth has been found and comprehended nor judge that it cannot be found and comprehended, but nevertheless persist, maintaining inquiry and contemplation."[10]

Gassendi's conflation of the goals of Academic and Pyrrhonian skepticism is not mere carelessness. It makes sense to equate the two uses of suspension of judgment given the polemical and entirely negative purposes of the *Exercitationes*. The *Exercitationes* endorses skeptical suspension of judgment as an anti-Aristotelian tactic without making any judgment about whether it is the end of inquiry or the proper response to the current state of inquiry.

[8] Brundell, *Pierre Gassendi*, 15 ff.
[9] Gassendi, British Library MS Harley 1677 (Book 8, Chapter 6), 37. The manuscript is numbered twice, once in pen and once in pencil. I always refer to the penciled numbers.
[10] Ibid.

Pierre Bayle wrote that before Gassendi, the views of Sextus were "no less unknown than the Terra Australis."[11] This is wildly exaggerated,[12] but Gassendi was instrumental in making Sextus' work better known. Gassendi understands Sextus as denying knowledge of essences or inherent qualities, but accepting knowledge of appearances so that we can, for instance, assent to the claim "honey tastes sweet to me" but not "honey is sweet in itself." For Gassendi, as for Sextus, appearances are not mental entities but ways external objects appear. Hence the presentation of an appearance implies that something that appears in a certain way is present to the senses and hence that something exists.[13] This comes out very clearly in his response to Descartes's skepticism about the external world. On Gassendi's account, we are immediately aware of things in the world as they appear to us. The main challenge for reconstructing his account of the truth of the appearances is to reconcile this direct realism with his denial of perceptual inconsistency or error, and beginning with his reconstruction of Epicurean canonic will help us meet this challenge.

Reconstructing Epicurean Canonic

Canonic is the first, and least developed, part of the tripartite Epicurean system of philosophy. According to Diogenes Laertius's remarks introducing Epicurus' "Letter to Herodotus," it "provides procedures for use in the system." The second part, physics, "comprises the entire study of nature," and the third and final part, ethics, "comprises the discussion of choice and avoidance." However, canonic was sometimes collapsed into physics, so that canonic became the part of physics that sets out the basic principles from which physical reasoning follows.[14] Canonic is sometimes called logic – and Gassendi's *Logica* provides a set of canons – but the Epicureans, at least according to Diogenes Laertius, distinguished canonic from dialectic and "reject[ed] dialectic as being irrelevant" on the grounds that "it is sufficient for natural philosophers to proceed according to the utterances made by the facts."[15] The rejection of dialectic is in the main a

[11] Bayle, *Historical and Critical Dictionary*, article "Pyrrho," note b.

[12] For an account of sixteenth- and seventeenth-century knowledge and use of Sextus before Gassendi, see Floridi, *Sextus Empiricus*.

[13] There is a great deal of secondary literature discussing whether the Pyrrhonian skeptics doubted the existence of an external world and, if not, why not. One well-known discussion is Burnyeat, "Idealism and Greek Philosophy."

[14] Diogenes, *Lives of Eminent Philosophers*, 10.29.

[15] Ibid., 10.31.

rejection of Stoic logic, although of course for Gassendi there is an echo of this in his rejection of Aristotelian logic.

Epicurean canonic has three criteria of truth: perceptions, preconceptions, and feelings. I discuss perception here and preconceptions – the ancestors of Gassendi's ideas – in Chapter 4. Feelings – in particular the feeling of pain (*molestas*) and pleasure (*voluptas*) – are the criteria for ethics and practical reasoning, and I omit them entirely.

To understand the way perception works as a criterion of truth, it is helpful to look first at the Epicurean account and then at Gassendi's reconstruction of it. Epicurus' insistence on sense as the main criterion is clear: "it is necessary," he says, "to observe all things in accordance with one's sense-perceptions... so that we can have some sign by which we may make inferences both about what awaits confirmation and about the non-evident."[16] For "if you quarrel with all your sense-perceptions you will have nothing to refer to in judging even those sense-perceptions which you claim are false."[17]

Maintaining this criterion requires denying that there could be any genuine conflict between the various deliverances of sense:

> A perception from one sense cannot refute another of the same type, because they are of equal strength; nor can a perception from one sense refute one from a different sense, because they do not judge the same objects. Nor indeed can reasoning [refute them], for all reasoning depends on the sense perceptions. Nor can one sense perception refute another, since we attend to them all. And the fact of our awareness of sense perceptions confirms the truth of the sense perceptions. And it is just as much a fact that we see and hear as that we feel pain; hence, it is from the apparent that we must infer about the non-evident.[18]

The perceptions of one sense-modality cannot conflict with the perceptions of another modality, for, as Lucretius explains, each sense has its own distinct jurisdiction.[19] Thus, every individual sense-impression is spoken of as "true": Both the round image received from a square tower seen in the distance and the square image received from that same tower close up are true. There is some scholarly dispute about what Epicurus and his followers mean by claiming that all sense-impressions are true.[20]

[16] Epicurus, "Letter to Herodotus" in Diogenes, *Lives of Eminent Philosophers*, 10.38.

[17] Ibid., 10.152–4.

[18] Ibid., 10.32.

[19] Lucretius, *De rerum natura*, 4.468–82.

[20] For a philosophically loaded interpretation on which the truth of the appearances bears a great deal of antiskeptical weight, see Everson, "Epicurus on the Truth of the Senses."

Fortunately, we can prescind from such disagreement because we only need to know what Gassendi took from Epicurus.

Gassendi endorses four Epicurean canons concerning the use of sense perception as a criterion. These four canons are a stable feature of his thought, appearing virtually unchanged from 1634 to the published *Syntagma*:[21]

Canon 1: Sense [*sensus*] is never deceived, and hence every sensation [*sensio*] and all Phantasies or appearances are true perceptions (1.53a).[22]

Sense cannot deceive, Gassendi explains, because all falsity is situated in affirmation or negation and sense does not affirm or deny anything "but only receives into itself *species* of sensible things and simply apprehends the thing that appears to it through species of it" (1.53a). When a square tower looks round, for instance, this is because sight has received a round *species* "for causes that have to be examined by physics" (1.53b). All the senses are equal in status, so one cannot correct the other. Nor can reason correct the senses on the basis of anything but consistency because reason depends for its function on the preconceptions previously acquired through sensory apprehension.

Thus, the senses, Gassendi writes, are the first criterion of truth and must be accepted, for two reasons. First, if we did not accept them, we would have no criterion of truth at all, and life would become impossible. Gassendi presumably has in mind here the objection that a true skeptic would refrain from judgment to such an extent that she would starve to death or walk off cliffs. The Pyrrhonians had detailed and sophisticated replies to this objection, but although Gassendi must have been familiar with them from the versions found in Sextus, he does not consider such replies. Second, Gassendi insists – as we have seen him do in the debate with Descartes – that we simply do not have any choice about whether to accept the senses: Some appearances simply compel belief.

Gassendi's second canon explains the source of truth and falsity:

Canon 2: Opinion, which follows the senses and is added to sensation, is that to which truth or falsity belongs (1.53b).

For more deflationary accounts, see Annas, *Hellenistic Philosophy of Mind*, 169–73; and Long, *Hellenistic Philosophy*, 21 ff.

[21] Gassendi, British Library MS Harley 1677. The four canons appear on pages 251, 268, 281, and 287, respectively. Gassendi also reiterates four canons of preconceptions (pages 294, 299, 303, and 308) and four canons of ethics (pages 312–13).

[22] I have benefited greatly from consulting Howard Jones's translations. Gassendi, *Pierre Gassendi's Institutio Logica (1658)*.

To judge that something is, or is not, as it appears is not the job of the senses but of the faculty of judgment, and hence truth and falsity belong to judgment. If we judge that the tower is round, this judgment, unlike the original perception, can be true or false. Gassendi gives two canons by which we can discover whether opinion is true or false:

Canon 3: That opinion is true which either supports or does not oppose the evidence of sense (1.54a).

Canon 4: That opinion is false which either opposes or does not support the evidence of sense (1.54a).

The notion of evidence at work here is psychological as well epistemological: One might just as well translate "evidence" (*evidentia*) as evidentness.

The claim that all sensations are true thus does not imply that all sensations provide equally good or accurate information about the world: Evident sensations provide better information. Gassendi writes that, for Epicurus, evident sensations are those "that – being free from all obstacles to judgment (such as movement, distance, the affection of the medium, and other things of that sort) – cannot be contested" (1.54a). The term "evident" is used to describe both perceptions and the objects they present, but the perceptual use is primary: An object is evident just in virtue of being presented by an evident perception.

Just as Gassendi takes Cartesian clear and distinct perception to play both a psychological and an epistemological role, for him evidence plays a dual role. He claims that sensory evidence – unlike Cartesian clear and distinct perception – can fill this dual role because the psychological compulsion it produces precedes and is presupposed by any further determination of truth, leaving no room for a criterion contradicting it. Psychologically, evident perceptions compel assent to what they present, and can be distinguished from nonevident perceptions on that basis alone. This is closely related to the claim that evident perceptions are those that are "free from all obstacles to judgment." For it is the existence of obstacles to judgment – or more precisely, things presented to sense that we take to be obstacles to judgment – that allows us to refrain from judgment on the basis of perception. When I perceive the distant tower as round, what allows me to withhold assent from the judgment that it *is* round is that I have previously judged that distance obscures vision. Thus, evidence is relative to the totality of the perceiver's perceptions, ideas, and judgments at that time.

Epistemologically, evidence plays the role of providing the first and foremost criterion of the truth of judgments. Gassendi's argument involves a number of strands, but the central theme is that there is nothing better than evidence to appeal to in making judgments, so that if we rejected evidence as a criterion, we would be unable to make any judgments at all. Moreover, because we cannot, on Gassendi's account, avoid assenting to evident impressions, he thinks there is no need to discuss whether evident perceptions could be misleading. Gassendi never brings up the possibility that human cognition could be misleading even when ideally used, but one imagines that he would give the same reply as Descartes, namely that such a possibility is inconsistent with the fact that we are creatures of a benevolent God.

The notions of support (or attestation) and opposition (or contestation) invoked in canons 3 and 4 rely on evidence. The evident perception of the tower from close up supports the opinion that the tower is square, that is, gives us reason to think that this opinion is true. It also contests the opinion that the tower is round. On the other hand, the perception of the tower from far away is not evident and thus does not either support or contest anything. Canons 3 and 4 are extremely weak: They allow us to count opinions that are not supported by any evidence at all as true so long as no evidence opposes them. Consider the two competing theories that the world is a plenum and that the world is composed of atoms moving in the void. The opinion that the world is a plenum, Gassendi argues, is contested by the evident perception of motion, and no evident perception supports it. On the other hand, nothing contests the opinion that there is a void. Hence atomism counts as true even *before* it is supported by the sort of evidence about crystalline structures, rarefaction and condensation, and chemical mixtures that Gassendi later adduces in its favor.

In the case of the void, Gassendi thinks, there is only one noncontested opinion, and hence only one opinion can be true. However, there are also cases where more than one opinion is noncontested. In such cases, Epicurus held that we should accept *all* the noncontested opinions.[23] His stock examples concern the heavenly bodies. As we have seen, Gassendi puts this Epicurean doctrine to use in the debate over heliocentrism. He holds that both Copernican and Tychonic theories are attested by a great deal of evidence and contested by none so that both are acceptable and the choice between them should be left to faith.

Evidence, then, plays both a psychological and an epistemological role for Gassendi. Indeed, his theory of perception as a whole draws a tight

[23] Epicurus, "Letter to Pythocles" in Diogenes, *Lives of Eminent Philosophers,* 10.87–8.

connection between the mechanics of sensation and the epistemic status of ideas and judgments. In this respect, it is very different from the scholastic accounts of sense perception he would have been familiar with, for those accounts were largely free of epistemological implications. Even the overarching claim that there is *nihil in intellectu nisi quod prius fuerit in sensu* (there is nothing in the intellect that was not previously in sense) was not typically taken to affect the epistemic status of intellectual beliefs. Aristotelians tended not to tie together questions about the origin of sensations, concepts, and beliefs with questions about their truth and accuracy in the systematic way Gassendi does.

So far we have seen Gassendi argue that all appearances are true because only judgments, properly speaking, can be false. This by itself, however, cannot do all the epistemological work Gassendi needs the truth of the appearances to do. He also holds that all appearances convey information and that examining the physical causes of the appearances provides the basis for knowledge of the world. This is why the truth of the appearances is the first and most important of Gassendi's canons. In order to see how he justifies this strong version of the claim that all appearances are true, we need to look at his account of the perceptual process and, in particular, of the objects of perception. I focus on vision because Gassendi discusses visual perception in far more detail than any other sense modality.

The Theory of Vision

Gassendi's account of visual perception distinguishes four things:

(1) The illuminated solid object, which has a certain atomic constitution or texture;
(2) The material *simulacrum* of that object;
(3) The impression created by that *simulacrum* in the sense-organ and then transmitted to the common sense, a material faculty located in the brain; and
(4) The mind's direct apprehension of the object of perception.

I discuss these in turn. A fifth entity is also relevant, although it is not a part of the perceptual process but a fortunate by-product:

(5) The material trace left in the brain as a result of the impression, a trace whose presence makes cognition of an absent object possible.

This fifth entity is sometimes called the *idea*, although the term is most often used for the act of cognizing an absent object and not the brain state that makes such cognition possible. Whether one thinks of ideas as mental acts or as the correlated dispositions of the brain, ideas are wholly irrelevant to the process of vision. They result from that process, but they need not be present for vision to occur. This point is crucial to understanding the version of direct realism that Gassendi has to offer, and it is one of the most notable ways in which Gassendi's account of perceptual cognition differs from the Lockean account it is often taken to resemble.

Let us begin at the beginning of the visual process. Gassendi uses the Epicurean term *simulacrum* to describe the entity intermediate between perceiver and thing perceived, and although his *simulacra* differ in kind from those of Epicurus, they play roughly the same role. Thus, it will be helpful to start with a quick look at Epicurus' account of vision. Epicurus held that because of their continual vibration, physical objects constantly throw off *simulacra*, films of atoms one atom thick.[24] Streams of *simulacra* travel through the air so quickly that they retain their coherence. When a stream of *simulacra* hits the eye, it rearranges its pattern of atomic motion, thereby imprinting an impression and disposing our eyes in such a way that we are caused to perceive those qualities possessed by the *simulacrum*. Gassendi explains that in the Epicurean theory, "the word *phantasia* is used . . . for that very appearance [*apparentia*] of the thing, or for that thing which appears as an apprehension both to sense and to the mind nudely (that is, without affirmation or negation)" (1.53b). This is, of course, completely different from the Aristotelian sense, in which *phantasia* is "that internal faculty that we generally call image-making" (1.78b).

Impressions, unlike the *simulacra* that cause them, do not have the properties of the object at the end of the causal chain. There is a sense in which a visual image of Socrates on the retina resembles Socrates: It has a similar shape, similar dispositions of parts, and so on. *Simulacra*, as copies of the surfaces of bodies, even more clearly resemble them. But the impression a *simulacrum* of Socrates makes is identified as an impression of Socrates because it is caused by him and not because of any resemblance.

Gassendi's first major amendment of the Epicurean theory is a change in the ontology of *simulacra*. He holds that *simulacra* are composed of rays of light, which in turn are composed of corpuscles and then atoms. Thus,

[24] Epicurus, "Letter to Herodotus" in Diogenes, *Lives of Eminent Philosophers*, 10.50.

a *simulacrum* is like the object that caused it in the sense that it provides the basis on which we can reconstruct a picture of that object, and not in the sense that the *simulacrum* itself is such a picture. This amendment is the result of several influences: the medieval perspectivist tradition, Kepler's recent work on vision, and Gassendi's own anatomical observations of the eye, made in a series of dissections of different animals in 1634–5 (*Mirrour* 4.67).[25]

Gassendi often cites the perspectivist tradition by endorsing the view that we "see under a pyramid" (2.381b). Although textbook writers such as Eustachius tend to speak as if intentional species are literal (albeit intentional rather than real) pictures, in more serious discussions the transmission of species is a short-hand way of talking about a two-dimensional image of the surface of an illuminated object re-created on the retina by rays of light transmitted from the various points on the surface of that object.[26] Gassendi more or less appropriates this account, making a few important changes. First, he emphasizes that the points from which rays of light flow are physical points – single atoms or single corpuscles – rather than mathematical points (2.378b). Although this claim does not make any real difference to vision, it is important for Gassendi's attempt to limit the intrusion of mathematical considerations into physics. Second, his atomistic account of qualities changes the status of the information received, as I shall explain later. Finally, Gassendi makes the changes necessary to accommodate Kepler's observation that the image on the retina is inverted (2.378b). Thus, he writes that

> [the most probable opinion] explains that vision occurs through the incursion of images that come in from the things themselves. In fact, rays of light come in from a variety of illuminated things and their illuminated parts and particles in such a way that they re-present those things individually [*ut eas singulas referant*]. And so the collection of rays that comes from one whole thing is the image or *simulacrum* of that thing, which image or *simulacrum* is represented like a thing in a mirror, so that it causes apprehension of this same thing in the sensorium. (2.377b)

Light-rays come from each point on the surface of an illuminated object in straight lines. These rays impact on the eye, and the ones that do so perpendicularly are sensed. This yields a correspondence between the surface of the object and the surface of the eye, so that an eye-sized

[25] For accounts of the first two, see Lindberg, *Theories of Vision from al-Kindi to Kepler*, 122–32 and 188–202. For the work done with Peiresc, see Bloch, *La philosophie de Gassendi*, 7–12.
[26] See Hatfield, "The Cognitive Faculties," 958 ff.

representation of the visible surface of the object is created on the surface of the eye. This representation is inverted between the surface of the eye and the retina.[27]

On Gassendi's account, the retina and the nerves that attach it to the brain are the most important parts of the visual system. For it is in the retina that the image that represents the accessible part of the surface of the object is transformed into information that can be conveyed by the nerves to the brain. Here, again, Gassendi alters Epicurus' account to make it consistent with seventeenth-century physiology:

> [A]lthough it seems to be the opinion of Epicurus that images or qualities of things go through the senses or external organs into the sense itself or the internal faculty and *anima*... nevertheless it is more probable that corpuscles slipping into the external senses do not penetrate into the interior faculty residing in the brain, but simply cause a motion of the nerves and the accessible spirits. The faculty residing in the brain is excited and moved by this motion stretching into the brain, and it apprehends that very quality of the thing in accordance with the impression of corpuscles, by means of an event like a mark (*factâ quasi notâ*). (2.339a)

Gassendi goes on to make similar alterations for the other four senses, although the mechanism of transmission is not worked out in as much detail. Hence, unlike the Epicurean account, which holds that sense perception takes place in the sense organs, on Gassendi's account sense perception takes place in the brain (2.371a–b).

How do we correlate the information received from the various different senses? It is clear that visual *simulacra* cause visual perception because the corpuscles of light that comprise them are able to interact with the structure of the optic nerves. Other *simulacra* thus must be constituted with different atomic structures, both to make sense of their different modes of transmission through the medium and to explain how they affect the particular sense organs they do. All *simulacra* have their effects in virtue of their corpuscularian structure, a structure that varies in type from organ to organ:

> [N]ot only the *simulacra* of colors, or images of visible things, but also sounds, odors, tastes and other qualities consist of corpuscles that are endowed with certain magnitudes, shapes, positions, and motions... the sense-organs – not only of vision but also of hearing, taste and the other senses – are constructed with spaces or pores or little passages that are... of a variety of magnitudes, shapes

[27] The *Animadversiones* gives the same account, save for this last step.

and positions that are commensurate to the various corpuscles, so that some can admit one sort of corpuscles into themselves, others another sort. (2.338a)

The corpuscles composing visual *simulacra* penetrate into the eye and hence move and affect the optic nerve. Similarly, the other four senses operate through the impact of corpuscles on the sense organ. This impact then causes a motion of the spirits in the nerves that in turn makes an impression in the brain. An impression is a change in the arrangement and motion of the different corpuscles that compose the brain:

[W]hen the external senses perceive their object, a certain motion occurs, first in the external *sensorium* itself, into which either *species* or qualities of sensible things come. Then, due to a certain propagation along the nerves, [a motion occurs] in the innermost brain where the nerves end. Then two things happen: one, the person perceives the sensible thing from which the blow came, which the sensing faculty has passed on to [the brain]; two, a certain vestige or as it were mark or image is impressed on the brain from this blow. (2.403b)

Gassendi describes three distinct events in this passage: the impact of the *simulacrum* on the sense organ; the transmission of that impact into the brain as an impression; and the act of apprehension that results from the impact. Both the language he uses and the details of the account suggest that content is not present until the last stage, the act of apprehension or perception. Impressions are that through which the mind apprehends content without being vehicles of content themselves. Content emerges only at the final stage, where the person "perceives the sensible thing from which the blow came."

This raises two questions. First, what *is* the sensible thing from which the blow came? In other words, what are Gassendi's objects of perception: ordinary solid objects, *simulacra*, or perhaps some third possibility? This issue is complicated by Gassendi's adherence to the Epicurean doctrine that the senses cannot lie. Second, how does the act of apprehension relate to the impression in the brain? This question has both a general and a more specific aspect. More generally, why does the impact of a pattern of corpuscles cause us to experience anything at all? In other words, how can something capable of apprehending content be constructed from insensate atoms?[28] More specifically, why does the impact of a particular arrangement of atoms cause us to apprehend one thing rather than another? How can we make sense of the particular connections between

[28] Despite his changing ontology of the mind, Gassendi remains committed to the view that sensation is a corporeal process shared by humans and animals alike and performed without the incorporeal soul.

atomic structure and content, even granting that the general problem is answered?

The Epicurean theory traditionally invoked a rather mysterious "fourth sort of atom" to explain how a sensate thing can arise from insensate atoms – or more accurately, in order to avoid having to explain this. The fourth sort of atoms is found only in the soul and is responsible for just those activities proper to the soul. We cannot sense atoms of the fourth sort or say too much about them, but they are not merely an ad hoc device. The claim is that sensation would not be possible if there were no such atoms, and hence the theory of signs legitimates accepting their existence.

While the fourth sort of atoms provides a way of dealing with the general problem of sensation, I am not aware of any Epicurean answer to the more specific question of how a particular arrangement of "soul-atoms" leads to apprehension of a particular content. Nor is it clear that the Epicureans really owe us an explanation of this. On their theory, we know that soul-atoms exist because without them, grasping content would be impossible – that is, on the basis of an inference to the only available explanation of a known phenomenon. To ask *which* dispositions of soul-atoms leads to *which* apprehension, however, goes beyond the limits of observation and inference to the best explanation.

Gassendi poses the general problem of sensation so that it applies to *any* theory on which minds with sensory capacities are composed of insensate parts. Epicurean atomism is, of course, the most salient example of such a theory:

How could sensate things, or things capable of sense, arise from insensible things? For instance, since [Epicurus] thinks that all things consist of atoms that are devoid of all sensation, how can it turn out that these things comprise an animal and the soul in the animal and a faculty or part of the substance from which the capability to sense is said to arise? (2.343a)

He provides the following answer on Epicurus' behalf:

[H]e does not, indeed, know precisely the magnitude, figure, motion, situation, and order of those atoms that are non-sensing in themselves but that nevertheless create sensing things when joined together. Nonetheless, non-sensing and inanimate things can be joined together in such a way that animate and sensing things result, just as non-igneous and non-hot things are joined together so that igneous and hot things result. (2.347b)

This is not much of an answer. Gassendi does not really respond to the original question but simply states that it is no more difficult for perceiving composites to arise from insensate matter than for macro-level qualities like color to arise from colorless atoms. He seems to imagine that just as atomism will one day explain light fully, it will explain the workings of the sensory power as well. However, he makes no attempt to give such an explanation. As we shall see when we look at the character of Gassendi's other explanations of macro-level qualities and powers, this does not put sensation in a category much different from any other purely macro-level qualities or powers. The atomism of the *Syntagma* is programmatic throughout.

It follows from Gassendi's account of perception that some mental operation is required to bring together the *simulacra* of the different senses. Gassendi is generally willing to talk about the mind having, say, the impression of Socrates imprinted on it as a result of perceiving Socrates. But in fact, there must be some further operation of the mind that intervenes between receiving the visual and auditory *simulacra* of Socrates and the formation of an idea of Socrates that brings in data from multiple senses. For when I see Socrates, this is a matter of my having a certain impression of vision; when I touch him, it is a matter of my having a certain impression of touch. Gassendi has little to say about how the impressions of the various senses are brought together, although this became a central issue for later empiricists like Locke and Berkeley.

It also follows from Gassendi's account that the mind must actively do something to consider one visual quality in isolation from other such qualities (for instance, to consider Socrates' color without considering his shape). Unlike Locke, for instance, Gassendi understands the first objects of perception as the totality of qualities of one sense rather than single qualities; we do not perceive Lockean simple ideas but must abstract them from the other aspects of the visual manifold presented to us. Thus, Gassendi holds that we are unable to formulate for ourselves a distinction between, for example, the black of Socrates' hair and his other visual qualities unless we already have an abstract idea of black available to us.

One final point is worth making. Gassendi's detailed accounts of perception and cognition stop where changes in the physical structure of the brain are made. He says almost nothing about how those changes relate to acts of apprehension. It is, in large part, this lack of information about the last step of the perceptual process that allows Gassendi to switch from the apparent materialism of the *Disquisitio* to the immaterialism of the *Syntagma* without making substantive changes in his account of cognition. I return to this issue in Chapter 10.

The Objects of Perception

Unlike later empiricists such as Locke and Hume, Gassendi emphasizes the mechanisms underlying perception rather than the nature of perceptual content. Thus, although understanding perceptual content is crucial for understanding his theory of cognition in general, we will need to do some reconstruction to see what Gassendi's objects of perception are.

There is substantial agreement in the early seventeenth century that substances themselves are not the objects of vision. The claim is something of a commonplace: Eustachius, for instance, insists that substances cannot be the objects of vision because per se substantiality is neither a sensible proper to any one sense modality nor a common sensible.[29] Dupleix concurs in rather stronger terms, writing that "we do not see or touch bodies, as the common people think, but only see their colors and touch their exterior surface."[30] For him, as for Eustachius, the sensibles are a certain subset of the accidents and properties of bodies. Thus, he takes the object of vision to be illuminated color, light and color being the two sensibles proper to vision. Descartes concurs with the tradition on this point, writing about, for instance "hearing, whose object is simply various vibrations in the ear."[31] In any case, it is clear from the example of the wax in the Second Meditation that we do not grasp substance through the senses.[32]

In a typical Aristotelian usage, the object of vision is something that we are directly aware of and that is present in the world (at least in cases where there is no distortion by the sense organs or the medium). Thus, the object of vision is that which is directly, although not immediately, perceived, where we perceive something *directly* if we perceive it without perceiving anything else and we perceive something *immediately* if we perceive it without using intermediate entities such as representations. A perception of the color of an external object is on this account direct, although it is mediated by an intentional species (on theories allowing for intentional species) or by light received in the retina (on theories dispensing with intentional species).

This way of thinking about the objects of vision fits in well with realism about sensible qualities. However, theories like Descartes's and Gassendi's

[29] Eustachius, *Summa philosophiae quadripartita*, Section 3.1.4.
[30] Dupleix, *La Physique*, 612.
[31] Descartes, *Principles of Philosophy*, 4.194 (French translation) / *Oeuvres de Descartes*, 9.314.
[32] Descartes, *Oeuvres de Descartes*, 7.30–2.

that do not construe colors as we are presented with them as real qualities of bodies cannot conceive of the objects of vision in this way. For the colors that I am conscious of do not really exist in the material world. Thus, for Descartes, the term "object of sense," no longer refers to that which is perceived directly: The relevant movement in the nerve is not perceived at all.

There are three constraints on understanding what Gassendian objects of perception are. First, the object of perception must also be the cause of perception. When Gassendi talks about the two ideas of the sun – one acquired through normal sense perception and the other through astronomical reasoning – it is clear that they are both ideas of the sun simply because they are both caused by the sun (3.316b–17a). However, it is also clear that in some meaningful sense they are *different* ideas: One is accurate; the other is not. Thus, Gassendi faces the problem, familiar to causal theorists of perception, of explaining how it is possible for us to have different ideas with the same cause. Like many causal theorists, he will give an answer in terms of the entire causal chain leading to the two ideas. Second, the account must be consistent with the claim that all appearances are true. The objects of perception must be such as to be represented accurately in all cases. Third, the account must accommodate the claim that one and the same object of perception can be presented more or less evidently.

The most natural reading of Gassendi's theory is that the object of perception is simply the ultimate cause, for instance the sun itself. However, this cannot be right. For one thing, it cannot explain why one perception of the sun is more evident than the other. Nor, more seriously, can it explain cases naturally described as cases of perceptual conflict. If we say that the content of the perception of a stick out of water and the same stick in water is the same, we will be left unable to explain why one looks different.

In the end, making sense of what Gassendian objects of perception are will require distinguishing what a perception is about from what it presents. Although this distinction is not made explicit in the text, it is the only way I can see to make sense of the claim that the appearances are always true, given what Gassendi wants to do with it. I shall explicate the distinction with reference to two cases where Gassendi provides extended discussion of a particular perceptual situation – the moon illusion and the case of color vision.

In *Epistolae quatuor de apparente magnitudine solis humilis et sublimis*, Gassendi discusses some optical illusions that arise in making astronomical observations and offers an atomistic explanation of them.[33] In doing so, he also provides grounds for explaining how optical illusions can occur when the appearances are always true. The chief illusion *De apparente* is concerned with is, in effect, the moon illusion, although Gassendi's discussion focuses on the sun and merely notes in passing that similar phenomena occur with the moon and stars.

Here is the problem. The sun appears larger to human observers when it is on the horizon than when it is overhead. Moreover, it creates a larger apparent diameter in any measuring apparatus used. Since the sun is presumably the same size and the same distance away in both cases, this is odd, and some explanation of the increase in apparent size is necessary. We cannot simply say that the two perceptions are of different objects because the same sun causes both. Nor can we simply say they are of the same object without explaining why different size images are presented in the two cases.

Gassendi's explanation of the illusion combines physical, physiological, and psychological factors. When the sun is on the horizon, the light it emits must pass through more of the atmosphere to reach us than when it is overhead because the angle at which the light traverses the atmosphere is more acute. Thus, the corpuscles of light emitted by the sun collide with more corpuscles of atmospheric vapor than they would at midday, and so the light is dimmer. Hence the pupils need to be dilated further, and so larger images are projected on the retina and the sun looks larger:

> [T]he sun appears larger when it is seen by the eye on the horizon than when it ascends higher because, when it is near the horizon, there is a long succession of vapors, and hence of corpuscles, that hold back the rays of the sun. Thus, the eye is not shut as far and the pupil, as it were enlarged by the shade, is much larger than when very rarefied vapors intercept the rays sent out from the sun and the sun shines so much that the pupil is much more contracted when looking at it. It seems to be for this reason that the visible *species* proceeding from the sun and sent into the retina through the amplified pupil occupies a more amplified seat in the retina, and therefore creates a larger appearance of the sun, than when the same thing reaches the faculty of sight via a contracted pupil. (3.421b)

One consequence of this account is that the stars should look larger in the middle of the night than they do at dusk and at dawn. So they do, Gassendi tells us; this is supposed to help confirm the account.

[33] For more on *De apparente*, see Chapter 6 of Joy, *Gassendi the Atomist.*

Gassendi's explanation is very strange. His claim is that the sun looks bigger on the horizon because the pupils are more dilated, and thus the sun projects a larger image on the retina. But if a greater dilation of pupils causes a larger apparent size, then surely *every* object should look bigger in dim light than in bright light. Moreover, if everything did look bigger in the dark because objects at a fixed distance occupy more space on the retina in dimmer light, then the visual field at night would have to be smaller. Neither phenomenon is generally recognized, and Gassendi does not embrace these consequences. However, he is discussing a situation where only one object, the sun, is represented in the visual field, and it is not after all so implausible to think that such situations work differently than everyday cases where different objects are juxtaposed in a way that gives us numerous contextual clues. Moreover, later letters of *De apparente* introduce a number of factors to consider beyond the size of the retinal image.

Whether or not the explanation is plausible, the most interesting feature for our purposes is its claim that the larger apparent size of the sun on the horizon is a fully accurate transcription of the information received on the retina. Thus, the sense of vision is not itself leading us into error; what vision provides leads to the sensation it should lead to. Of course, we would be wrong if we judged that the sun actually shrank and grew as the day progressed. However, we have no difficulty refraining from such judgments, presumably because other experience suggests that real change in size is unlikely.

The sun itself, as represented by the light-corpuscles it sends to the brain, as altered in transmission, and as received by a human eye in a given context, fully determines the size of the image projected on the retina. This image in turn fully determines the impression created in the brain. Thus, Gassendi explains the apparent size of the sun as the result of a confluence of three factors: the sensed object itself, the way the *simulacrum* is altered in transit, and the disposition of the eye.[34]

This explanation suggests understanding the claim that the appearances are true as the claim that they accurately represent their causes. If we take the object of perception to be simply the object at the end of the causal chain (for instance, the sun itself), then the two appearances

[34] Bloch, *La philosophie de Gassendi*, 11, points out that the *Syntagma* deemphasizes the material train of transmission and emphasizes what goes on in the nerves. This does not affect the moral I draw from the example.

of the sun as of different sizes cannot both be true. If we take the object of perception to be simply the *simulacrum* as it impacts the sense-organ, then both appearances are true – but now they seem to be appearances of two entirely different things. Thus, we need a sort of duality of content, a way of understanding the object of perception as both the distal and the immediate cause.

Given such a duality of content, it is relatively easy to see how some perceptions can be more evident than others although all are true: More evident perceptions convey more information about their distal causes and less information about the medium and the perceptual apparatus. Gassendi's overarching claim is that the appearances "must appear in the way they do owing to their causes" and that these causes should be investigated by physics (3.388b ff). The point of such physical investigation is to separate perceptual information about ordinary objects from perceptual information about other parts of the causal chain.

More help understanding Gassendian objects of perceptions can be derived from his account of color. Colors are not fundamental constituents of things because individual atoms have no color. Gassendi entertains three possible ontologies of color: (1) colors may be real qualities arising from the texture of macroscopic bodies, just as the capacity to sense arises from certain textures of insensible atoms; (2) colors may be powers to affect the human perceptual system that arise from textures; or (3) colors may be appearances (1.432a). He rejects (1) because it fails to explain the light dependence of colors, maintaining the common view that things do not have color in the dark.[35] However, in various places he suggests both (2) and (3). Hence I shall offer a reading on which their disagreement is merely terminological.

Gassendi consistently claims that different "dispositions" or arrangements of surface corpuscles affect the light diffused by the body in different ways. Powers arise from such arrangements in the presence of light. And he consistently claims that light diffused in different ways causes different impressions on the brain and hence perception of different colors. It is somewhat unfortunate that the term "color" is used for both the power and the appearance, but, if I am correct that Gassendi has in mind two distinct aspects of perceptual content, it is unsurprising.

We can think of color vision as depending on two factors: the body seen and the effect of the medium and mechanism of perception. Take a pair of perceivers, both of whom are looking at a white wall. One sees it as

[35] Guerlac, "Can There Be Colors in the Dark?"

white and the other as yellow, for reasons of medium or mechanism. The distal cause of their two perceptions is the same, and in that sense they perceive the same wall. However, what is presented is different because somewhere along the way the causal chains diverged, resulting in two different impressions.

What guarantees that the two different appearances are both true is that both accord perfectly with the impression made on the brain; indeed, on Gassendi's account, appearances cannot help but accord with impressions. Of course, one perceiver probably has a tendency to make the false judgment that the wall is itself yellow. As Gassendi reminds us, we must be careful not to leap to the conclusion that things in themselves resemble their appearances. But the possibility of false judgment gives us no reason for concern about relying on the senses. It merely serves as a warning that we must understand how sense perception works before rushing to judgment on the basis of sensory appearances.

This reconstruction places no weight on what soon came to be called the primary quality–secondary quality distinction. Although Gassendi does distinguish qualities like size and shape that pertain to atoms as well as composites from qualities like colors, that distinction does not map onto the distinction between appearance and reality in any neat way.

Gassendi's account of the objects of perception, as I have reconstructed it, distinguishes what perceptions are about from the appearance they present. Noticing this distinction helps us make sense of Gassendi's assumption that we can suspend judgment about the natures and intrinsic qualities of things but not their existence. When I am presented with an appearance, I am guaranteed that it has some real external cause. I have no guarantee, however, that what is presented to me corresponds to that distal cause rather than to the effects of the medium and mechanism of perception. The reading I offer has the virtue of explaining how Gassendi can both hold that the doctrine of the truth of the appearances is epistemically significant and claim that we can doubt only the nature and qualities, not the existence, of the things we perceive.

Gassendi's theory of sense perception and the related accounts of ideas and judgment are very closely tied to an explanation of the source of error and the way to avoid it. The connections between what we now call psychology and epistemology are much closer for Gassendi than they were for typical scholastics, perhaps because of his concern with the conditions under which skeptical suspension of judgment is possible. Consider, for example, Gassendi's psychologistic notion of evidence: The opposite of evidence is obscurity, which results when we see something from so great a distance or in such bad conditions that we cannot make out its components clearly. Gassendi's revival of Epicureanism reintroduces considerations about the human perceptual and cognitive apparatus into epistemology and scientific method.

4

Cognition, Knowledge, and the Theory of Signs

Gassendi is an empiricist of a more radical form than his scholastic predecessors. In the tradition he was most familiar with, the process of abstracting from sensible species to intellectual species was often held to add new content: the essence, which is not contained within the sensible species. Gassendi, like other early moderns, keeps the term "abstraction," but for him abstraction is a process of cutting away content rather than creating new content. Hence it cannot enable us to grasp mind-independent universals or essences.

Gassendi enshrines the claims that all ideas originate from sense and that all content is acquired through sense in the fifth canon of his reconstructed Epicurean logic:

> *Canon V*: Everything that is an anticipation or praenotion [that is, idea] in the mind depends on sensations, either by incursion or proportion or similitude. (1.54b)

There are no ideas innate in us, merely an "innate faculty of knowing" that allows us to acquire ideas through the senses and then construct new ideas from them (3.379a). Most of Gassendi's logic is devoted to explaining how this innate faculty of knowing constructs adequate ideas of things we do not sense on the basis of what's acquired through the senses. But as well as providing this account of simple apprehension, Gassendian logic also includes judgment or proposition, by which ideas are combined into sentential complexes, and reasoning.[1] And because reasoning aids in the

[1] The fourth branch of logic, "Method," is devoted to teaching rather than discovery, and hence I omit it entirely.

formation of new ideas – in particular, ideas of unobservable entities that can be used to explain the appearances – understanding where content comes from requires looking at all three phases of cognition.

The Origin of Ideas

As we have seen, perceptual cognition does not make use of ideas but yields them as a fortunate by-product. When *simulacra* impress themselves on the sense organs, this produces an impression in the brain that leaves in its wake a "trace" or disposition of the corpuscular structure of the brain. This trace provides the basis for cognition of absent objects, just as impressions provide the basis for perceptual cognition. Gassendi sometimes refers to this brain trace as an idea and sometimes uses the term 'idea' for the act of cognition dependent on that trace, so that the term 'idea' has both an occurrent and a dispositional sense. Other more or less synonymous terms such as anticipation, apprehension, conception, form, image, *notio anticipata,* phantasm, and representation tend to preserve this duality.[2]

Of course, we can think about objects we have never sensed, and hence Gassendi cannot hold that all ideas copy impressions. Instead, he adopts three Epicurean ways of forming new ideas: through *enlargement and diminution,* as "when from the idea of a man of normal size the mind gets the idea of a giant by enlarging, or from the same idea gets the idea of a pygmy by diminution"; through *unification,* as "when from the ideas of a man and a horse [we] get the idea of a centaur"; and by *analogy,* as "when the mind transfers and adapts the idea of a city one has seen to a city one has not yet seen and thus pictures the latter as being like the former" (1.93a).

The first two methods are straightforward, so long as we confine ourselves to the causal and imagistic terms that Gassendi uses to model the mind in the *Institutio logica.* However, Gassendi does not think that ideas are literally images in the mind any more than he thinks that impressions are literally images in the brain. Thus, it is clear how enlargement and unification work in terms of content, but it is much less clear how they work in terms of physiology. Because Gassendi follows the typical structure of scholastic texts and discusses idea-formation and reasoning in the

[2] These uses are all found in the Preface of the *Institutio* and the introduction to Part I (1.91a–2b).

context of logic rather than underlying physiology, he never provides a physiological account of these processes.[3]

Analogy raises issues in terms of content as well as physiology. For analogy is supposed to produce the ideas of a number of things that are centrally important to Gassendi's philosophy: atoms, God, and the immaterial human soul. None of these entities are very clearly analogous to anything we sense. Atoms lack most of the sensible qualities that the things we have sensory ideas have, thus making the analogical idea of an atom – such as, for instance, a very small billiard ball – misleading. And it is hard to see how immaterial entities like God and the immaterial human soul bear much analogy to anything sensible. Gassendi tells us that our idea of God is formed by analogy with "some grand old man" (1.93a), but this hardly provides all the content he needs.

As well as explaining the formation of ideas of particular objects we have not sensed, Gassendi also needs to explain how ideas of *types* of things are formed. He offers two alternate methods: aggregation, which produces what I shall call general ideas, and abstraction, which produces what I shall call abstract ideas.[4] A new idea is formed from previously acquired particular ideas by aggregation when

> the mind as it were separates out similar ideas and makes one aggregate from them, which, since it contains all [the individuals], is the idea of all of them taken together and hence is called universal, common or general . . . [for example,] the aggregate of the ideas of Socrates, Plato, Aristotle, and all the other similar things, which, because the name *man* is commonly given to these singulars, is generally called *the genus man.* (1.93a)

The aggregate represents the objects of all the particular ideas from which it was constructed. Aggregate ideas have intentionality in exactly the same way the particular ideas from which they are formed do. Just as the idea of Socrates represents him because he was the first step in the causal chain, the aggregate idea of man represents him because he is part of its cause.

However, the aggregate idea must also represent other men of whom the thinker has had no experience. Gassendi's account of abstraction

[3] It is clear that the processes of enlargement, diminution, and unification are carried out in the brain by the corporeal soul and not by the incorporeal soul. For, as we shall see in Chapter 10, when Gassendi assigns cognitive labor to the immaterial soul, it includes only such higher-level processes as self-reflection and cognition of universals.

[4] Gassendi distinguishes the two types both in this section of the *Institutio logica* and in the *Disquisitio* (3.379a).

relies on the existence of objective similarities between members of the same species to explain this. When I group the particular ideas of Socrates, Plato, and the like together, I am guided by similarities among them, for those similarities typically lead to similarities among their ideas. When I meet another man and form an idea of him, it will in normal cases be immediately clear that he is similar to the men already encountered (1.93a–4a).

These similarities and the mental mechanisms that grasp them also explain how we can distinguish those qualities of Socrates that constitute his being a man from irrelevant qualities like his snub nose. For making such a distinction is not supposed to involve reasoning or the application of further ideas. We do not group Socrates and Plato together after recognizing that they have certain features in common; rather, we are able to note the qualities in respect of which they are similar only *after* we have formed the general idea *man*. The process of grouping ideas of similar things together in the mind happens mechanically or automatically – it is a matter of impressions of similar things as it were stacking themselves on top of each other. Exactly how this is supposed to work is unclear, but it is important for seeing the motivations of the story, and its differences from its Lockean cousin, that on Gassendi's account it takes a great deal of effort for us to break the world down into the simple qualities that constitute it.

How do what I am calling abstract ideas differ from general ideas? The general idea of man represents many incompatible features of men (having blond hair, having black hair, and so on) because it contains the ideas of particular men. The genuinely abstract idea of man prescinds from all such differences between individuals:

> it is given that these singulars are similar in some ways or are mutually agreeing but nevertheless have many differences by which they differ from one another. Therefore the mind looks at them separately and thus as it were separates out all those things in which all the singulars agree and takes away or does not look at the differences by which they differ. So it has, as a common, universal . . . idea, that which has been abstracted and contains nothing not common to all. This is what is called the genus. For example, when the mind notes that the ideas of these same Socrates, Plato and Aristotle are in agreement and similar in that they represent a two-legged animal, standing erect, reasoning, laughing and capable of being taught, it separates out these features (i.e. two-legged . . . [etc.]) and forms an idea from which all the differences have been omitted (such as the fact that one is the son of Sophroniscus, . . . [etc.]). And it takes such an idea as the . . . universal idea of man, insofar as it represents not this or that or another particular man, but man in general or in common. (1.93a–b)

It is in considering the status of such abstract ideas that the ontology of the mind is most relevant. Gassendi casually remarks, concerning abstract ideas, that

> it is difficult, not to say impossible, to distinctly imagine man in general so that he is neither tall nor short nor average in height, neither old nor young nor in his middle years, neither black nor white nor any other special color. But we should at least keep in mind that a man whom we intend to represent in general should be free from all these discriminations. (1.95b)

Later sections of the *Syntagma* suggest that the mind's immateriality is proven by its ability to possess genuinely abstract ideas (2.441a, 2.251b). In light of that claim, this passage might be thought to imply a need for an immaterial mind. However, Gassendi does not spell out *why* only an immaterial mind could possess abstract ideas, so it is not clear that the difficulty of imagining an indeterminate idea is the reason. Moreover, in the *Disquisitio* Gassendi seems to think that a mind conceived as material can operate with genuinely abstract ideas.

Thus, I suggest that the difficulty Gassendi alludes to is really a difficulty in making clear and distinct ideas that prescind from most of what was evident in perception. It is not entirely clear why it would be any more difficult for a corporeal idea to represent man in general with a nose but no particular nose than to represent insensible qualities. In the former case, sense provides too much information, too many causally relevant noses; in the latter, it provides too little. The first problem seems much easier to deal with because it simply requires a mechanism for cutting down on sensory information. Only excessive devotion to the metaphor of acts of cognition as looking at ideas in the brain – a metaphor Gassendi relies on quite heavily but cannot intend literally – could make the representational ability of truly abstract ideas into a special problem.

Gassendi's account of the acquisition of aggregate and abstract ideas implies the same order of cognition as standard scholastic textbooks.[5] We cognize substance after accidents; universals after singulars; the more universal after the less universal; and the parts or *differentia* of complex substances after the substances themselves (since distinct cognition is posterior to confused cognition).[6] This agreement is deliberate

[5] See, for instance, Eustachius, *Summa philosophiae quadripartita*, 427–9.

[6] Substance after accident because we sense qualities but judge that they are qualities of a substance, universals after singulars because ideas of universals derive from ideas of singulars, and the more universal after the less universal for the same reason.

and programmatic. Gassendi's chief objection to Aristotelian accounts of cognition is not that they enumerate the possible objects of cognition incorrectly or get the order in which our cognitive capacities develop wrong (although we shall see one case where he does object in this way).[7] Rather, he objects to the underlying metaphysics.

The Quality of Ideas

The quality of ideas is a matter of their clarity and distinctness, in the following sense. Ideas are better or worse in proportion to how many of the parts and attributes of their object that they represent (1.95a–b). Thus, the idea of a particular man is more complete if it represents, say, the peculiarities of his face. Gassendi notes that "it is for this reason that we are led to recommend anatomy, chemistry and the other sciences, which distinguish and uncover for us more parts and qualities of a thing than are normally apparent" (3.386b).

Objective reality is, for Gassendi, a matter of the quality of ideas. An idea has objective reality to the degree it represents the object it is about (the distal cause) or, to put it another way, to the degree that the object it presents (the appearance) is similar to the distal cause. Thus, Gassendi objects to the causal principle Descartes uses in the first argument for God's existence in the Third Meditation:

> the idea, or its objective reality, should not be measured by the total formal reality of the thing (i.e. the reality that the thing has in itself) but merely by that part of the thing of which the intellect has acquired knowledge (i.e. by the knowledge that the intellect has of the thing). Thus you will be said to have a perfect idea of a man if you have looked at him carefully and often from all sides; but your idea will be imperfect if you have merely seen him in passing. (3.323a)

It follows that we acquire ideas gradually. The ordinary sensory idea of man is less evident than the anatomist's idea, but they are both ideas of the same thing because they ultimately derive from the same cause, namely, men. Gassendi counts this gradualism as an advantage of his account over Descartes's and others on which ideas have to be acquired all at once because the forms or essences they involve are single and indivisible.[8]

7 This contrasts with Gassendi's objections to Descartes's account of cognition: Descartes, he argues, *has* failed to account for the available data about the order of cognition. For instance, on Descartes's account we could have an idea of substance in general before we had ideas of particular substances, but experience shows that this never happens.

8 For the significance of this claim, and for textual evidence on Descartes's side, see Lennon, "Pandora, or, Essence and Reference," 162.

For experience, he says, shows that we do acquire ideas gradually:

> I . . . can have an idea of a kind of animal never seen by you. How do you show that this idea is what you call an innate idea? . . . you will never attain knowledge until it is described or pointed out by someone. However, when your eyes have shown it to you, you derive an image of it . . . and then at length you realize that you have an idea of it in you. But beyond this, how will you know that something else, already provided in your mind, is excited? . . . Nothing that is true and already there is excited, as is shown by the fact that when we see only one individual of a certain kind of animal, we know by experience that we have . . . no idea other than the kind that is derived from this sort of individual. . . . Moreover – to show that a certain idea is not innate, but is made by us from observations of individuals, there is also this fact: the idea is perfected through time and becomes increasingly more general insofar as we know more things and prescind from many observed differences. But if it were innate or even excited after seeing a single individual, it would come forth absolutely perfect and not always change or stand in need of eliciting. (3.379a)

This particular point fails to affect Descartes's position because Gassendi misunderstands which Cartesian ideas are supposed to be innate. However, the general point stands: Nativist theories that hold that ideas may need to be excited by experience but are not derived from experience have no good explanation of why repeated experience improves our ideas.

Both types of general ideas can represent more or less well, depending on how refined they are. Aggregate ideas are better when the collection of things they represent contains more examples: The idea of man that includes native Americans is more complete than the one that includes only Europeans and Africans (1.95b). Abstract ideas are more perfect insofar as they contain more of the features that are common to all the relevant similar particulars, and *only* those features. An abstract idea of man that did not allow us to see that all men stand upright would be imperfect, as would one depicting men in general as snubnosed.

On Gassendi's account, ideas represent things in virtue of being caused by them. More precisely, ideas represent those parts of things that were copied by the *simulacra* that form the basis of the idea. Given this, it is is not difficult to see how ideas can represent the *physical* parts of things more or less accurately. However, it is more difficult to see how attributes or "metaphysical parts" can be represented (1.98a–b). The idea of a man, Gassendi says, is more complete the more clearly it represents his "talent, memory, virtue, wisdom and so on" (1.95a). It is unclear how something conveyed to one of the five senses through *simulacra* could be invoked in order to explain how the idea of a particular man represents his middling virtue and bad memory. Gassendi offers no causal mechanism that could convey those properties to the mind.

In his account of ideas Gassendi is trying to bring together scholastic and Epicurean *desiderata*, and although the attempt is interesting and important, it is not entirely successful. From the Epicurean side, Gassendi argues that the representational capacity of ideas must ultimately be explained in terms of physical mechanisms of transmission. From the scholastic side, he asks for the representation of an object to depict its metaphysical structure, and it is hard to see how an Epicurean account of concept acquisition can yield such representations.

This is a general problem with no clear answer in Gassendi's work, although we shall see later that he does have a clear answer to one specific form of the problem. The general problem is how a causal and largely materialist account of representation can account for the representation of insensible or incorporeal properties such as virtue. This problem is closely entwined with imagism. Gassendi does not in the end think that ideas are pictures or that apprehension is a matter of looking at mental pictures, but the metaphors he uses suggest an underlying imagist intuition. Thus, at a first pass, we might pose the general question by asking how an imagistic theory could account for the representation of insensible qualities.

It is, indeed, hard to see how insensible qualities can be represented, on Gassendi's view. But posing the problem in terms of imagism locates the difficulty in the wrong place – inside the mind rather than in the causal chain from world to mind. Thus, it suggests that there is no problem understanding how modes of the brain can represent sensible qualities. But it is merely the metaphor of mental vision that suggests we can explain how sensible qualities are represented in entirely subject-internal terms. The difficulty understanding how imperceptible qualities can be represented is really a difficulty understanding how representations of imperceptible qualities can be formed on the basis of what the senses receive. Hence, we should be able to solve the problem by giving a fuller causal story. Providing such stories is a central task of Locke's and Hume's accounts of human understanding.

A more specific form of the problem is how such a theory of representation can account for the location of ideas in the network of genus and species, and Gassendi has a relatively clear answer to this: Representation is holistic. Maximally clear and distinct ideas contain various relations that the object of the idea has to other objects. A paradigmatic example is relations to things up and down the Porphyrean tree from it:

The idea of a thing establishes its division into species, parts and attributes. For whenever an idea represents a thing as a genus, it represents it as containing species. Whenever an idea represents a thing as a whole, it represents it as consisting of parts. And whenever it represents it as a subject, it represents it as the subject of attributes. Accordingly, the more or less imperfect the idea is, the more or less imperfect a division of genus into species, whole into parts, subject into attributes, can be established. (1.97b–8a, italics his)

The general idea of a horse, for instance, represents individual particulars as horses, and thus represents *horse* as a genus with different species falling under it.[9] It is helpful to recall the complaint in the *Exercitationes* that on the Aristotelian account of genus and difference, we would have intimate knowledge of everything if we knew the *differentia* of the least thing (3.184b). The tree-structure of genus and difference works holistically. Gassendi's account shares this holism while avoiding its unfortunate consequence because, on his view, the clarity and distinctness of ideas comes in degrees. It may be true that the perfectly clear and distinct idea of, say, a chicken would require us to have clear and distinct ideas of everything else along that branch of the Porphyrean tree. But a usable idea of a chicken just requires having seen a few different animals and thus having derived the genus *animal*.

Similarly, sufficiently clear and distinct particular ideas, like general ideas, contain within themselves information about the relations their objects bear to other things:

From the idea of a thing, its relation to other things is understood. It is understood from the idea of each individual thing not only what it is like in itself or absolutely, but also what it is in respect to other things or comparatively. So . . . from the idea of man we understand not only that in itself and absolutely man is a rational animal, but also that man is a genus in respect to species, a whole in respect to parts, a subject in respect to accidents. (1.98b)

If we have even a minimally clear and distinct idea of man, we know that *man is an animal*. Thus, we learn the *definitions* or *principles* of the objects of our ideas at the same time as we acquire reasonably clear and distinct ideas. For we acquire knowledge of definitions just in virtue of having the experiences necessary to form a sufficiently clear and distinct idea. As has been noted, there is no clear distinction in Gassendi's theory between acquiring ideas and making judgments.[10] When we attend to sufficiently clear ideas, our attention automatically produces

9 The lowest species are individuals.
10 Ayers, *Locke*, 15.

definitions and axioms. An account of judgment falls out of the theory of ideas.

Judgments can also emerge from comparing ideas. Gassendi seems to think that all comparisons are in respect of similarity:

> *Canon VII*: An anticipation is the principle in all reasoning; for, attending to it, we infer that one thing is the same, or different, conjoined, or distinct, from another.

Just as when I consider the ideas of Socrates and Plato, I see that they are similar enough to count as ideas of men; when I compare the idea of a particular man with the abstract idea of animal, I see that the first is sufficiently similar to the second to count it as the idea of an animal. When this process works properly, the comparison issues in a true proposition because the similarity in ideas derives from similarity in the world. When it does not work properly, false propositions result because the ideas are similar, but their objects (the things that caused them) are dissimilar or at least similar in the wrong respects. If I form an idea of animal that includes being a quadriped, then I might falsely judge that men are not animals because my idea had failed to capture relevant similarities between bipeds and quadripeds. All false judgments result from some misrepresentation in the ideas comprising the judgment.

Syllogistic Reasoning

Gassendi allows for two different types of reasoning: syllogistic reasoning and the sort of reasoning that pertains to inferences from signs. Typically, syllogistic reasoning is demonstrative, and significatory reasoning is merely probable, but the distinction between the two types is not clean. Syllogistic reasoning can be probable as well as demonstrative, if the premises are probable rather than certain, and all inferences from signs use a syllogism as one of their component parts. For the sake of clarity, however, I shall begin with syllogism.

For Gassendi, all reasoning is discursive: There is no Platonic or Cartesian intuition of truths. Intuition can yield judgments by showing us that two ideas agree or disagree, but it cannot connect different judgments. To make such a connection, there must be some idea shared by both judgments so that it can be used as a middle term. This way of thinking about reasoning issues in the view that all reasoning is syllogistic in form because it always involves two propositions sharing a middle term and leading to a conclusion:[11]

> when the mind apprehends that two notions coincide with a third by means of a twofold act of proposing, it immediately brings them together and proposes that they coincide with each other. . . . And thus a syllogism is nothing but thinking or internal speech that infers a third proposition with necessity from two previously proposed propositions. (1.106a)

Consider an example that includes all three operations of the intellect:

> someone sees a body a mile away. He apprehends it with the intellect, but proposes nothing about it, such as whether it is animate or inanimate. He looks attentively and observes that it changes place or progresses, and consequently apprehends motion or progression. So two simple apprehensions, one of body and the other of progression, are accomplished. At once, a judgment or second operation follows, in which the intellect, connecting the two first simple apprehensions, judges or pronounces that the body is progressing. He does not stop at this, however. Because he has formed the judgment or second operation *that which progresses is an animal* from previously held notions, he repeats the first judgment and puts it next to the new one, and connects them so that a third judgment is elicited, in the following manner: *that which progresses is an animal; that body is progressing; therefore that body is an animal.* And this is reasoning, or the third operation of the intellect, which is also a judgment, but a judgment that depends on others and is more composed. (3.367a–b)

Even a very simple chain of thought, when spelled out fully, takes the form of a syllogism. It is this insistence on the discursivity of reason that underlies the *Disquisitio*'s puzzling tendency to point out alleged weaknesses in Descartes's arguments by first recasting them in syllogistic form.

Gassendi explains the necessity of discursivity in the course of discussing enthymemes in the *Institutio logica*:

> when an enthymeme is called an imperfect syllogism, this has to be construed only as far as verbal expression is concerned. So far as the mind is concerned it is perfect. . . . If the premise suppressed in verbal expression was not grasped by the mind, the mind would not feel or undergo the force of inference from whence it adds the conclusion. Indeed, when someone says, for example, *every animal senses, therefore man senses,* he infers the conclusion only because he also understands that man is an animal. . . . if this did not happen, there would not even be an imperfect syllogism but a random inference, as if anything at all followed from anything else, for example, as if one said *the moon is in the sky, therefore man senses.* (1.112a)

[11] There is an exception for the case of what was called "immediate inference," that is, inferring that *no S is p* from *no p is S.* However, on Gassendi's view this is best seen not as reasoning but as another sort of judgment because no new idea is introduced.

Without discursive, syllogistic form, Gassendi argues, inferences would simply be random transitions with no logical force. Thoughts do not always come into the mind by orderly sequences of syllogisms; we can think one moment that the moon is in the sky and the next that man senses. But this is not reasoning at all, not even bad reasoning. It is simply the arbitrary succession of thoughts.

Gassendi recognizes that the operations of the mind can, in introspection, appear to be immediate. The judgment that a moving body in the distance is an animal will typically appear immediate to introspection for judgments are formed from simple apprehensions so quickly that in many cases we do not notice their formation:

> whatever is known about something is contained in the idea of it. From this it follows that the more things the mind has ideas of, the more abundant and extended its *scientia* is, and the more things the singular ideas contain clearly and distinctly, the more splendid its *scientia* is. Thus, in a man of great wisdom, *scientia* is almost without discourse and is like simple understanding, which, intuiting the idea, looks over the antecedents and consequences in one view. But in a more unsophisticated man, there is no *scientia* without considerable difficult discourse, because he needs speculation and delay in order, once the antecedents have been known, to see their consequences. (1.99b)

The speed at which judgments are formed from ideas depends largely on how good the relevant ideas are. The ideas of the wise are so clear and distinct that the element of discourse happens too quickly to be noticed: As soon as they bring the idea to mind, they intuit the true propositions to be derived from it.

Reasoning and the Theory of Signs

A Pyrrhonian question asks for the criteria of legitimacy for any reasoning that goes beyond the ideas derived from evident perceptions and finds no acceptable criteria. The Gassendi of the *Exercitationes* and a great deal of the *Disquisitio* seems to grant this: Our knowledge of appearances can never lead to knowledge of essences. But in more constructive moods in the *Disquisitio*, and in much of the *Syntagma*, Gassendi does allow inference from appearance to underlying reality. He enshrines such inferences in the last of the canons of Epicurean logic that I shall discuss:[12]

[12] The remaining canons are divided into two groups, one concerning ethics, where pleasure and pain are the ultimate criteria, and the other concerning the use of words, where Gassendi recommends clarity and simplicity in discourse. I am not aware of any work on

Canon VIII: What is not evident should be demonstrated from an anticipation [or idea] of an evident thing.

Such demonstrations are inferences from signs and play a crucial role in knowledge of unobservable entities. An inference from signs has three elements: a perceived object, property, or event *x*; the idea of an unperceived object *y*; and a premise, usually left implicit, to the effect that *x* can only exist when *y* exists or that *y*'s existence explains the existence of *x*. I call this third element the bridge premise.

There are three different categories of unperceived object (1.68b–9a):

(1) The altogether unperceived or totally hidden, that is, things that can neither be perceived themselves nor signified by any perceived things. Gassendi gives the strange but traditional example of the oddness or evenness of the number of stars.

(2) The unperceived by nature, that is, things that cannot be directly perceived but can be known by inference from what is perceived. These come to be known using *indicative signs*. Standard examples are atoms and the void; the patterns of motion of the heavenly bodies; and (before the invention of the microscope) the pores in the skin.

(3) The unperceived at the time, that is, things that are not currently perceived but that could be perceived at some other time or place. Such things can be known using *admonitory* (*commonefactivum*) signs, as when a scar signifies a wound, smoke fire, or lactation pregnancy. After the invention of the microscope, the pores in the skin became an admonitory sign.

This tripartite distinction and the related distinction between indicative and admonitory signs played a central role in Hellenistic debates about knowledge. The distinction between indicative and admonitory signs is originally a medical one. The Empiricist doctors claimed that admonitory signs could give us all we needed and that indicative signs should be rejected, while the Rationalists defended the use of indicative signs.[13]

Not much attention was paid to justifying the use of admonitory signs, for, as Gassendi notes, even Sextus accepted that inference from an effect

the canons involving language, but there is a helpful discussion of the role of pleasure and pain as ethical criteria in Sarasohn, "The Ethical and Political Philosophy of Pierre Gassendi."

[13] Allen, *Inference from Signs*, 107 ff.

to its temporarily hidden cause is allowable.[14] The fact that we can later perceive something previously known only through an admonitory sign helps justify the use of such signs. Moreover, it has been suggested that we should understand the bridge premise as something supplied by memory, so that strictly speaking an inference from admonitory signs is memory rather than reasoning and thus not open to skeptical critique.[15]

The use of indicative signs relies on the premise stating that the sign could not exist unless the signified also exists. In the case of things hidden by nature, memory cannot supply the bridge principle because things hidden by nature are by definition things we have never perceived. Thus, we need some other source for the bridge principle. Consider a famous Epicurean example of reasoning through indicative signs: The void must exist because motion could not exist unless void existed. Thus, motion is a sign of the void. How do we come to know that motion is impossible without an absolute void? Epicurus' answer is that we know this by analogy with evident things. In ordinary cases, a (relatively) solid body cannot move somewhere unless that place is (relatively) empty.

The legitimacy of this particular use of analogy is open to question, as is the legitimacy of analogical justification in general. One Stoic criticism of Epicureanism is that the use of analogy relies on an assumption that the unperceived is just like the perceived – which assumption is unwarranted, as well as incompatible with Epicurean atomism. The most detailed ancient discussion is Philodemus's *De Signis*, rediscovered only well after Gassendi's death. Philodemus summarizes a Stoic objection to the method of analogy as follows: "According to the method of analogy one ought to infer that since all bodies in our experience have color, and atoms are bodies, atoms have color; or since all bodies in our experience are destructible, and atoms are bodies, atoms must be admitted to be destructible."[16] To this, Philodemus replies that *of course* one must be careful using analogy. We cannot legitimately analogize from "any chance common qualities," but only from qualities always possessed by all the evident objects we use as the basis of analogy. The argument that atoms have color is illegitimate because it fails to recognize that not all bodies have color. Similarly, not all bodies have been observed to be destructible, so analogy does not suggest that atoms are destructible. In contrast, the inference from motion to a void is legitimate because *all* evident bodies in

[14] Sextus, *Outlines of Scepticism*, 93 (Section 2.10.102).
[15] Allen, *Inference from Signs*, 110 ff.
[16] Philodemus, *Philodemus: On Methods of Inference*, 33 (Section 5).

all circumstances need a relatively empty space to move into.[17] Philodemus also gives several other relatively commonsensical rules about the use of analogy. It is crucial that we need both the support of evident perceptions for the principle assumed in making the analogy and an absence of contestation of that principle by other evident perceptions. Only drawing analogies in cases where both these conditions are met is supposed to rule out those analogies that would create problems for atomism and still allow those that justify it. Gassendi could not, of course, have been familiar with Philodemus's reply to the skeptic, but the analogies he uses typically fit both these constraints.

Gassendi's significative inferences, like the Epicurean ones, typically rely on a principle bridging from the macroscopic to the microscopic world and hence on the assumption that microscopic bodies behave like macroscopic bodies. However, this assumption causes problems for Gassendi. For, as we shall see, Gassendi's universe is radically different at the atomic and macroscopic levels. Bodies have sensible and occult qualities but atoms have neither. Bodies are always divisible but atoms cannot be divided. Worse, atomic motion turns out to be radically different from the motion of composite bodies for Gassendi grafts a Galilean account of the motion of composite bodies onto a roughly Epicurean account of atomic motion. This threatens the legitimacy of the analogy between atomic and microscopic motion that forms the basis of Gassendi's main argument for atomism.

It is thus difficult to see how Gassendi can maintain that principles bridging from the level of bodies to the level of atoms are evident. Indeed, it is often difficult to see what would even license thinking of them as probable. However, in certain cases, it is clear that he does think they are evident: That no two bodies can be in the same place at the same time, for instance, is evident. Gassendi justifies this by suggesting that we can have evident perceptions of relations between ideas as well as of external objects. Certain perceptions of relations between ideas are not contested:

> [D]emonstration is not used and is not necessary when matters are so evident that merely describing them does the work, as is the case not only with particular things that are open to sense and established by experience, but also with general propositions against which no objection can be given, as is the case with the axioms to which mathematicians reduce their demonstrations, such as *the whole is greater than its parts.* (1.86a)

[17] Ibid., 65 (Section 17).

Gassendi's tactic is to assimilate certain logical truths to the evidence of sense perception. If this tactic works, it renders the bridge premises of inferences from signs legitimate, at least when those premises can be construed as relations among ideas.

Gassendi is here trying to draw a parallel between the evidence of sense and the evidence of reason. But what exactly does the parallel consist in? Evidence comprises a number of features, but the most relevant one seems to be that evident perceptions compel belief. This gives us a way of picking out our evident perceptions. Similarly, Gassendi might say, in some cases we simply cannot help but believe what reason tells us. We cannot, as a matter of psychological fact, suspend judgment on all matters. Presumably, *two bodies cannot be in the same place at the same time* is supposed to be irresistible in this way.

In inferences from signs, the bridge premise can be evident to reason, and the sign itself can be perceived with evidence. Thus, one might think that Gassendi should hold that the *conclusions* of inferences from signs can be evident as well. However, he does not, so far as I know, ever describe the conclusions of chains of reasoning as evident, and there are two reasons why he might not allow that the conclusions of significative inferences are evident.

We have seen that for a perception (including the perception of a relation between ideas) to be evident implies that we are certain its content is true and that we are compelled to believe it. We are only rarely, if ever, compelled to believe things on the basis of reasoning. Even though reasoning tends to bring about belief in its conclusions, such belief is rarely inescapable. Gassendi's frequent reminders about the limitations of human reason, for instance, can function to limit belief. Thus, one reason Gassendi may be unwilling to claim that significative inference can produce evident conclusions is simply his overarching emphasis on the weakness of human reason. He is willing to admit that simple propositions like *two bodies cannot be in the same place at the same time* are perceived with evidence, but he is not willing to extend such evidence to the sort of significant natural philosophical conclusions inferences from signs are supposed to issue in.

Second, in the vast majority of cases, the conclusions of significative inferences are probable rather than certain and hence could not possibly be perceived with evidence. We have seen examples of significative inferences whose conclusions are supposed to be certain, although not evident: the inference from sweat to pores, for instance, or from motion to the void. In these cases, the sign can only exist in one circumstance.

However, most significative inferences involve competing explanations, and we have to adjudicate between those explanations in order to pick the most probable.

The distinction between appearances that could only signify one underlying state of affairs and appearances that could signify a number of things is one Gassendi makes repeatedly. In 1642, he wrote to de Valois that

> One must attend to and discern between those things that are consonant with the appearances in a singular way, and those that are consonant in many ways with what happens to us.... an example of the first sort is the existence of the void and the indivisibility of the atoms. (6.156a)

A common example of the second is the motion of the heavenly bodies. This example must have been irresistible for Gassendi because the example is licensed by his Epicurean predecessors and enables him to discuss legitimately both Copernican and Tychonic hypotheses in astronomy.

It is useful to compare the status of such dual hypotheses to cases where Gassendi offers corpuscularian explanations of phenomena like magnetism or the deadly glance of the basilisk. If these phenomena are real, on Gassendi's view, they must be explained by the underlying corpuscularian structure. For instance, the effects of the magnet must be explained by some corpuscular emission from the magnet to the attracted body. So the inference from magnetic effects to the emission of some sort of corpuscle is a case where we can envisage only one possible underlying cause, and hence it counts as certain. But exactly what the corpuscular emissions are like is a matter of mere probability, in exactly the same way that we have mere probability about the structure of the heavens. In the next chapter, we shall see a case where significative inference yields a unique, certain conclusion – atomism – and then a number of cases where multiple probable explanations are proffered.

5

Space and Time

Gassendi bases natural philosophy on three principles, as many Aristotelian philosophers did.[1] However, for the three Aristotelian principles matter, form, and privation, he substitutes space, time, and the atoms that constitute material bodies and whose motion constitutes their causality. The claim that space exists independently of bodies is necessary for Gassendi's atomism: Atoms move through the void, so independent space was a feature of both Democritean and Epicurean atomism. Gassendi also holds that time has a "remarkable parallelism" (1.133a) with space, and hence construes time as well as space as having an absolute existence independent from bodies or their motion. This is rather more novel, and it is nowhere near as clear that Gassendi's atomism requires absolute time. I begin by discussing space, time, and the void; in the next chapter, I move on to the world and the atoms that compose it.

Space, Place, and Vacuum

The terms space, place, and vacuum or void were widely used in ancient and medieval philosophy. Aristotle takes place as the primary notion and defines void as "place with nothing in it." Because he concludes that the void is impossible, space and place end up being the same thing.[2] Gassendi reverses the direction of argument, first arguing for the possibility and existence of space distinct from the place of any body and

[1] See for instance, Colegio das Artes (Coimbra, Portugal), *Commentarii Colegii Conimbricensis Societatis Iesu, in octo libros Physicorum Aristotelis,* 184 (L1, C, Q1, A4).

[2] Aristotle, *Physics* 4.7, 213b32–14a3.

then for the existence of void space. The argument for the distinctness of space from bodily place depends on a claim that God could eliminate bodies and their places without eliminating the space within which they are located.

At this point, the question whether there is actually void space arises. Gassendi conceives of three ways in which there could be a void: there might be an *extramundane* void, that is, a void outside the world; there might be an *interstitial* void, that is, a void spread throughout the interstices within matter in the world; and there might also be macro-level *coacervate* voids, that is, accumulated voids created by mechanical means (1.186a). Gassendi argues that all three types exist. An extramundane void must exist because of the ontology of space. The interstitial void must exist because it is what makes possible the motion evident to sense. In contrast, Gassendi's argument for the coacervate void is a posteriori and relies mainly on his interpretation of the barometer experiments. I discuss the arguments concerning the three sorts of void, the properties of space that emerge from these arguments, and the thorny issue of the relation between space and God.

The issue of whether a vacuum was possible and, if so, whether any such void exists in nature has a long and much-studied history. We may as well begin with Aristotle's denial of the void and his claim that motion could not occur in a void. Against Democritean atomism, Aristotle argued that the speed of bodies moving in a void could not be explained and hence that motion through a void would be impossible.[3] In his discussion of space, Patrizi notes that Aristotle's argument applies only to motion in an extra-mundane void.[4] Macroscopic bodies moving in only relatively void spaces within the world *do* encounter some resistance. However, the same problem recurs at the level of atoms because individual atoms can move without encountering resistance – as they do whenever they do not collide with another atom.

A fundamental assumption of Aristotle's argument is that the time a body takes to move a certain distance is proportional to the resistance offered by the medium.[5] From this it would seem to follow that all bodies

[3] Aristotle also argued that the *direction* of bodies moving in a void could not be explained. However, this argument relies on a distinction between natural and violent motion that Gassendi rejects and hence I shall leave it out. See *Physics* 4.8, 215a1–23.

[4] Hence it would be irrelevant to Patrizi's claims because Patrizi does not think anything moves in the void spaces beyond the world. Patrizi, *Nova de Universis Philosophia*, 63b. Compare Henry, "Void Space," 143.

[5] Aristotle, *Physics* 4.8, 215a24–16a11.

moving in the void move with the same infinite speed, which is absurd. Aristotle's Epicurean opponents accepted that all bodies move with the same speed in the void – and hence that atoms, which only move through void, always move with the same speed – but denied that this speed is infinite.[6] A second Aristotelian argument against the void also involves motion but is quite different in character. Assuming, as the atomists do, that "void is a sort of place deprived of body," then motion will be impossible because void cannot penetrate body, bodies being impenetrable.[7] This argument introduces two themes that become central in later discussions: the role of impenetrability in the nature of body and the question of whether bodily dimensions and spatial dimensions can be distinguished.

Neither of Aristotle's arguments are intended to show that the void is impossible – merely that if it existed, motion through it would be impossible. Late medievals tended to abandon the claim that the void was altogether impossible, and so it became more or less taken for granted that a void is conceptually and logically possible, if not naturally possible. Seventeenth-century textbooks illustrate this. Eustachius, for instance, states that no vacuum occurs naturally in this universe, although God could create a vacuum. Moreover, he insists, *contra* Aristotle, that motion could most likely occur in a vacuum. Aristotle is wrong to think that the resistance of the medium is the only relevant cause of succession of motion: one should also bear in mind "other causes," such as "the distance of the end-points between which motion occurs."[8] Eustachius's discussion – which continues by insisting on the sheer implausibility of the claim that a man in a vacuum could not move his finger – need not concern us. All that matters is his quick dismissal of the problem of motion in a vacuum.

The possibility of the void was also disputed on purely a priori grounds, again originally by Aristotle and then by a long series of others. It is worth quoting Aristotle's somewhat obscure argument at length:

> even if we suppose [the existence of a void settled], the question of what it is presents difficulty – whether it is some sort of 'bulk' of body or some entity other than that; for we must first determine its genus.
>
> Now it has three dimensions, length, breadth, depth, the dimensions by which all body is bounded. But the place cannot *be* body; for if it were there would be two bodies in the same place.[9]

[6] Lucretius, *De rerum natura* 2.225–29; compare 6.325–47.
[7] Aristotle, *Physics* 4.8, 214a18–23; compare 216a27–b2.
[8] Eustachius, *Summa philosophiae quadripartita*, 92 (Section 1.3.2.5).
[9] Aristotle, *Physics*, 4.1, 209a3–8; compare Aristotle, *Metaphysics*, 13.2.

If this argument succeeded, it would rule out both extra-mundane and interstitial voids, for both are supposed to be three-dimensional. However, it is puzzling why we are supposed to grant that extension is body once we grant that it is three-dimensional. Indeed, Gassendi seems to think we can dissolve this problem just by noticing that matter is by nature three-dimensional and impenetrable but space is by nature three-dimensional and penetrable.

Descartes, who is perhaps the most famous heir of this argument, fills in the gap with his thesis that the essence of body is extension. Thus, a void, a region of extension containing no body, is conceptually impossible. The denial of the void is a feature of Descartes's system from early on. In the autumn of 1631, he wrote to Mersenne that although "everyone imagines that there is" a void, "one cannot suppose a vacuum without error."[10] The familiar argument that "it is just as impossible that a space should be empty as that a mountain should be without a valley" is found by 1638.[11] This argument recurs at *Principles* 2.18, where Descartes attempts to correct the false belief of childhood that voids exist in nature on the grounds that "nothingness cannot possess any extension."[12]

Aristotle's equation of three-dimensional void space with body was found in a number of scholastic philosophers as well as in Descartes. However, it was firmly rejected by the Italian natural philosophers Patrizi, Telesio, and Campanella. Patrizi, whose 1591 *Nova de Universis Philosophia* gives an account of space very similar to Gassendi's, is the most important. Indeed, Gassendi admits his debt to Patrizi on this matter, writing that "concerning space or place . . . Patrizi presented nothing other than the reasoning presented earlier" in the *Syntagma* (1.246a). References to Patrizi are common in the *Syntagma*; however, favorable ones like this are not: Generally Patrizi, one of the "obscure philosophers" and an adherent of the theologically unacceptable doctrine of the world soul, is an opponent.

Patrizi disputes both the claim that motion would be impossible in a void and the claim that any three-dimensional extension is a body. Distinguishing physical space from mathematical space, he argues that God first created physical space and then created the matter in it. This is one of the major points on which Gassendi, who takes space to be

[10] Descartes, *Oeuvres de Descartes*, 1.228. The context makes clear that Descartes thinks everyone imagines that a vacuum is *possible*, not that everyone imagines that a void actually exists.

[11] To Mersenne, November 15, 1638. Descartes, *Oeuvres de Descartes*, 2.440.

[12] Ibid., 8.50.

a condition on the possibility of creation rather than a created thing, disagrees with him. Another point of disagreement is over what the void is devoid *of*: For Gassendi, the definition of void as space without matter implies that there is *nothing* in void space, but for Patrizi the void is filled with incorporeal light.

Patrizi argues that space must be prior to matter, "for it is necessary that what exists before all other things is that thing by which, when supposed, all other things can be supposed and by which, when taken away, all other things are destroyed."[13] Here he affirms in very similar words what Aristotle had used the argument quoted earlier to deny.[14] Patrizi's space is characterized by its three-dimensionality, its immobility, and its penetrability or lack of resistance. Body, on the other hand, is determinate three-dimensionality together with *antitypia* or resistance. The three-dimensionality of space is a key feature in Patrizi's attack on the Aristotelian conception of space, which attack consists mainly in casting doubt on the notion that a two-dimensional thing like the superficies or surface of a body could contain a three-dimensional thing like a body.[15]

Much of Patrizi's discussion of space is devoted to explaining its ontological status. We have to say that space is *aliquid*, an *ens*, to avoid saying that space is nothing at all and hence making unclear why removing space would remove all bodies.[16] But what sort of thing is space? Since it "subsists *per se*, inhering in nothing" it cannot be an accident.[17] Since space was created first, before any of the bodies in it, it cannot be quantity: Quantity, like the other eight categories excepting substance, depends on substance. Patrizi is more tempted by the possibility that space is a substance:

> If substance is that which subsists *per se*, then space will be a substance in the highest degree. If substance is that which exists *per se*, then space will be a substance in the highest degree.... If substance is that which other things subsist in, then space will be a substance in the highest degree.... If substance is that which is first of all beings, then space will be a substance in the highest degree.[18]

However, Patrizi is reluctant to endorse the view that space is a substance without qualification, since space does not fit the *category*

[13] Patrizi, *Nova de Universis Philosophia*, 61a.
[14] Aristotle, *Physics* 4.1, 208b23–9a2.
[15] Patrizi, *Nova de Universis Philosophia*, 61b–3a.
[16] Ibid., 62a.
[17] Ibid., 65d.
[18] Ibid.

substance very well. Space is not an individual substance because it is not a matter-form composite, and it is not a genus because neither species nor individuals are predicated of it.[19] Thus, Patrizi, at a bit of a loss about what to call space, labels it "an incorporeal body, and a corporeal non-body."[20]

Once space is given this unique ontological status and is no longer dependent on substance, it is clear that void space should be possible. In fact, Patrizi discusses the same three types of void as Gassendi, although he does not introduce a distinct term for the sort produced by various machines.[21] He provides a strikingly atomistic analogy to explicate the notion of an interparticulate void. Just as there are tiny particles of air between the grains of sand on a beach, there are small portions of void space between those particles, and so on for other macroscopic bodies. Rarefaction and contraction result from increase and decrease in the amount of void space between particles. Patrizi also suggests, like the Epicureans and Gassendi, that such tiny voids are necessary for motion.[22] (He simply ignores the Aristotelian arguments against motion in a void.) These atomistic suggestions are not developed elsewhere in *Nova de Universis Philosophia*, and Patrizi's model of matter emanating from light and, indirectly, from God suggests a continuous structure.[23]

Patrizi gives a rather more a prioristic – and not, on its own, terribly satisfying – argument for the extramundane void. The stars, he says, are bodies and hence have superficies, dimensions, and sizes; moreover, there are distances between then. Therefore there must be space in the heavens. This space must be (actually) infinite, since it cannot be limited by other bodies, other spaces or an incorporeal entity.[24]

Acceptance of the void was not limited to the neo-Platonist tradition. Throughout the 1640s, for instance, Hobbes accepted not merely the possibility but the actuality of a void within nature. The 1642–3 *Anti-White*, for instance, admits the possibility of a vacuum.[25] And in 1648,

[19] Ibid.

[20] Ibid.

[21] Patrizi's discussion, like Gassendi's, refers to various machines – like water-clocks, siphons, automata, self-trimming lamps, and organs blown by windmills – from Hero of Alexandria's *Pneumatics*.

[22] Patrizi, *Nova de Universis Philosophia*, 63a–c.

[23] This is, of course, merely a suggestion and not an implication. See Henry, "Void Space," who notes that there are also atomist elements in Patrizi's account of mathematical space, which relies on indivisibles.

[24] Patrizi, *Nova de Universis Philosophia*, 64a–d.

[25] Hobbes, *Thomas White's De Mundo Examined*, 47–9 (Section 3.9).

after reading Père Etienne Noël's response to Pascal in "Le Plein du Vuide," Hobbes wrote to Mersenne: "I still think what I told you before: that there are certain minimal spaces here and there, in which there is no body. . . . Certain small empty spaces are necessarily formed by [the action of heat-producing bodies]."[26] He and Gassendi were in agreement on this, as on the further point that the various barometer experiments "do not prove that a vacuum exists."[27] (Later, Hobbes was to decide that there was in fact a fluid ether that played the role of the void without being absolutely nothing.[28]) I return to the barometer experiments and their varying interpretations later. My goal here has been simply to show some of the breadth of options available on the nature of space and the chief arguments against the void that Gassendi had to contend with.

The Ontology of Space

Gassendi thought seriously about the nature of space very early on in his career, perhaps because this was a question on which the Italian natural philosophers he had read disagreed violently with scholastic Aristotelianism. Gassendi was not yet committed to atomism in 1631, but he was interested in it, and this may have piqued his interest in the nature of space. In a letter to Peiresc of 28 April 1631 that was printed as a preface to the *Exercitationes*, Gassendi wrote that he planned to reinstate the space of the ancients and dispense with Aristotelian place, reestablish the void in nature, and treat time "differently" from Aristotle (3.102). The plan for the never-written Book 3 is carried out in the *Animadversiones* and the *Syntagma*, whose atomism requires a space empty of bodies and thus distinct from the Aristotelian place that is the place of body.

One of Gassendi's first concerns in laying out his new principles of natural philosophy is to reject the Aristotelian ontology of place. In Aristotelian ontology, every existing thing – every *res* or *ens* – is either a substance or an accident. Gassendi's account of space relies on rejecting this principle entirely. In this he is following Patrizi. However, Gassendi's

[26] Hobbes, *The Correspondence*, 2.164–5.

[27] Ibid., 1.172–3.

[28] The fluid ether was also supposed to play the role of Descartes's subtle matter. In February 1657, Hobbes told Samuel Sorbière: "I did not think that Epicurus' theory was absurd, in the sense in which I think he understood the vacuum. For I believe that he called 'vacuum' what Descartes called 'subtle matter', and what I call 'extremely pure ethereal substance', of which no part is an atom, and each part is divisible . . . into further divisible parts" (Ibid., 1.444–5).

account of the ontology of space is far more precise and detailed than Patrizi's.

Gassendi describes space and time as *res verae* (true things) and *entia realia* (real entities) that are nevertheless not substances (1.179a). Within the tradition, there are two criteria of substantiality: Substance is that which exists per se, and that which is the subject of accidents (1.179a–80a). Space fails to meet the second criterion, Gassendi argues, for accidents do not subsist in space but in particular bodies. (Brownness, for example, does not inhere in space but in the cow.) But space *does* meet the first criterion since, as we shall see in a moment, space exists per se and thus cannot be an accident of bodies. Thus, space satisfies one criterion of substantiality but not the other, and hence requires an ontological category of its own. Gassendi revises the Aristotelian principle that *every being is either substance or accident* to say instead that "[e]very Being is either Substance, or Accident, or Place in which all Substances and all accidents are, or Time in which all substances and all accidents endure" (1.182a). Space and time are entities *sui generis.*

One might wonder what justifies rejecting the Aristotelian ontological scheme simply on the basis of an argument that space fits neither the category of substance nor the category of accident. It was quite common to put space under other categories. For instance, Dupleix takes place to be a sort of quantity.[29] Sennert lists a number of options, including accident but also quantity.[30] Eustachius is somewhat more precise, holding that in one sense place "pertains to the category *ubi* [where]" and in another that it is "a species of continuous quantity."[31] However, in the *Exercitationes* Gassendi took himself to have established that the eight categories over and above substance and accident are fictions of the mind. Relying on this previous argument, Gassendi assumes that if we accept the categories, then we must either think that space is a substance or an accident, or else admit that it is merely fictional. And he takes it as evident that space and time do have some mind-independent reality, remarking that they "do not depend on the intellect like a chimera because whether or not the intellect thinks, space endures steadfastly and time flows on" (1.182a). Gassendi gives no argument for this claim, but none is needed in this context. The view he is trying to defeat is not one on which space and time have no reality, but rather one on which their reality is dependent

29 Dupleix, *La Physique*, 249–50.
30 Sennert, *Epitome Naturalis Scientiae*, 92.
31 Eustachius, *Summa philosophiae quadripartita*, 86.

on that of substance. The ten categories are not supposed to be *mere* ways of thinking about the world, with no foundation in reality. So it is fair to think that if the opponent can claim no more than that space and time are real as real conceptions in the mind, then the battle is already lost.

The significance of Gassendi's new ontology of space has been widely discussed, especially in the context of the development of Newton's view of space.[32] By placing space and time outside the categories, Gassendi rejects Aristotelianism far more radically than his main competitor, Descartes. Indeed, insofar as Descartes identifies space with body and time with the duration of substance, Descartes's conception of space and time is roughly Aristotelian.[33] The Cartesian conception of space was rejected by both Newton and Leibniz, and both were familiar with the Gassendist alternative.[34] In his mature work, Newton arrived at conceptions of space and time that had much in common with Gassendi's.

Gassendi was not, of course, the only one of his contemporaries to reject the Aristotelian and Cartesian conceptions of space. Consider Henry More, who is also sometimes thought to have influenced Newton's conception of space.[35] More described space, in language similar to that used by Gassendi and Patrizi, as an "infinite immobile extension"[36] that is "[o]ne, simple, immobile, eternal, complete, independent, existing from itself, subsisting by itself, incorruptible, necessary, immense, uncreated, uncircumscribed, incomprehensible, omnipresent, incorporeal, permeating and encompassing everything."[37] These are terms that also apply to God, on More's view, and indeed More thinks of space as "a certain rough representation of the divine essence."[38] However, unlike Gassendi, who makes it clear that space and time are in the same ontological category, More has almost nothing to say about time.

[32] For famous early discussions, see Koyré, *From the Closed World to the Infinite Universe* or Koyré, *Newtonian Studies.*

[33] For space, see Descartes, *Principles of Philosophy*, 2.9–12 / *Oeuvres de Descartes* 8.7-9; for time, see Descartes, *Principles of Philosophy*, 1.57 / *Oeuvres de Descartes* 8.27.

[34] For Newton's objections to the Cartesian conception of space, see "De gravitatione" in Newton, *Philosophical Writings*, 14 ff. For Leibniz's objections, see the "Conversation of Philarete and Ariste" in *Die philosophischen Schriften von Gottfried Wilhelm Leibniz*, 6.584–5. For Leibniz's familiarity with Gassendi, see Konrad Möll, *Der junge Leibniz*, 2.29 ff. Newton's familiarity may have been via Charleton's *Physiologia Epicuro–Gassendo–Charletoniana.*

[35] For a survey of arguments for and against this suggestion, see Hall, *Henry More*, 206 ff.

[36] More, *Henry More's Manual of Metaphysics*, 54.

[37] Ibid., 57.

[38] Ibid., 61.

Absolute Space and the Extra-mundane Void

Gassendi takes the logical and conceptual possibility of a void space existing outside the created world for granted.[39] His strategy is to argue from the absolute possibility of an extramundane void to its possibility within the actual order of nature and finally to its actuality. To do so, he uses of a version of the annihilation experiments found in Patrizi and quite common in late scholasticism.[40]

The argument begins with the following thought experiment, intended to show the existence of incorporeal dimensions. Consider the lunar sphere. Imagine that God has destroyed "the entire mass of elements within it" so that there is now nothing within the lunar sphere:

> I ask whether, after this reduction to nothingness has occurred, we still conceive the same regions that had existed between the concave surfaces of the lunar heavens but are now empty of elements and void of all body? Certainly, nobody could deny that God could preserve this lunar heaven and reduce the bodies contained in it to nothingness and prevent any other bodies from going into their place unless they were to deny the power of God. (1.182b)

Gassendi then argues that the mere possibility of coherently supposing a void beneath the lunar sphere shows that there are incorporeal dimensions:

> [B]ecause the lunar heaven is an orb, can't we, taking one point on its concave surface, conceive that there is a certain interval or distance from it to the opposite point? And isn't this distance a certain length – say an incorporeal and invisible line – that is the diameter of the region [under the lunar heaven] . . . ? In particular, don't the dimensions of length, breadth and depth that we had imagined still remain there? Undoubtedly, whenever a certain interval or distance can be conceived, a dimension can also be conceived because there is a determinate measure of such an interval or distance, that is, they can be measured. Therefore, the dimensions that we said were incorporeal and spatial are of this sort. (1.183a)

39 Someone following Grant, *Much Ado About Nothing*, 108–10, might explain this assumption as a consequence of the condemnation of 1277. Grant argues this had far-reaching effects, leading to a greater emphasis on God's absolute power (that is, his power to do anything that is not a logical contradiction) and an insistence that God's causal power is not limited by the principles of natural philosophy. Thus, the 1277 condemnation brought about consensus that an extra-mundane void is possible, in the sense of being within God's absolute power, even if it contradicts the principles of natural philosophy and the laws of nature.

40 Leijenhorst, *The Mechanisation of Aristotelianism*, 109 ff.

The argument can be extended to show the possibility of *infinite* void as well. Assume, Gassendi says, that "a world that was bigger and bigger to the point of infinity had previously existed" (1.183a). That is, imagine an infinite lunar sphere. Now imagine that God had annihilated the matter under that infinite sphere and then re-created it, just as before. Three consequences follow from this second thought experiment. The first is that space itself is totally immobile – it remains when everything movable within it has been moved out of the way – and the second is that its dimensions are incorporeal. I discuss these two at greater length below. The most important consequence, however, is

> that space would be boundless [*immensa*] before God created the universe; that it would still be there if God were perchance to destroy the world; and that of his own free will God chose that determinate region of space in which to create the world (leaving the residual space, generally called *Imaginary* space, all around it)... that it would still remain if He should destroy the universe, and that of his own free will God chose this determined part of space's realm in which he created the universe (leaving the remaining space, generally called imaginary space, around it on all sides). (1.183a)

This amounts to the claim that there could be immeasurably vast space existing independently of the world. And because it is not up to God to actualize this possibility, it must always have been actual.[41]

Gassendi has thus given an argument, relying on a notion of God's power that puts aside the constraints of the principles of physics, for the existence of incorporeal dimensions and hence the distinction between corporeal and incorporeal dimensions. He has other defenses of the distinction between corporeal and incorporeal dimensions as well. He invokes the precedent of Nemesius, who had insisted that although "every body is endowed with three dimensions... not everything endowed with three dimensions is a body."[42] More compellingly, he defends the distinction between corporeal and incorporeal dimensions by claiming that resistance or impenetrability is a criterion of body, but not of dimensionality per se.

[41] At this point, one might wonder how Gassendi has ruled out the possibility that God must *first* create incorporeal dimensions and *then* create a world within them. Gassendi's answer here is simply that incorporeal dimensions are not the sort of thing that can be created because they are not positive entities. It seems that few readers find this compelling.

[42] Gassendi quotes Nemesius's *De natura hominis,* Chapter 2, at 1.182b. Gassendi was familiar with his *De natura hominis* because it contained a survey of theories of the soul, including the Epicurean theory.

On Gassendi's view, it follows, from the possibility of incorporeal dimensions and the distinction between corporeal and incorporeal dimensions, that incorporeal dimensions and corporeal dimensions are distinct even when they coincide. Thus, there must actually be incorporeal dimensions wherever there is body: If the dimensions of a place can exist when that place is empty, then the two sorts of dimension are distinct, and so the incorporeal dimensions must still exist when the place is full. At this final step, where Gassendi asserts the actuality of incorporeal dimensions, he has moved from a discussion of what is consistent with God's absolute power to a discussion of the world God actually created.

Gassendi does not make the move from God's absolute power to the actual order of nature entirely clear. Margaret Osler has argued that such moves should be understood in the context of his strong voluntarist tendencies: Because God puts no limits on what he can actually do within the order of nature by establishing that order, what is possible in terms of God's absolute power is possible within the order of nature as well.[43] However, we need not see any such motive operating here. If it is true that the existence of independent incorporeal dimensions is required for God to exercise his absolute power, then those incorporeal dimensions must continue to exist *after* God has established the order of nature. The key is that Gassendi does not say that God could *create* incorporeal dimensions. This would not show that he had actually created such dimensions. Rather, Gassendi's claim is that what God can do, even setting aside the constraints of the order of nature, requires that there *already are* incorporeal dimensions independent of his actions. This is equivalent to saying that there is space distinct from the place of body – that is, absolute space – even within the world.

Gassendi thinks this same line of argument also shows that there are infinite or unbounded incorporeal dimensions as well. Remember that we were supposed to grant that God could create a lunar sphere that was infinite in dimension. Now, if it is really possible for God to do this, then it must be possible for there to be infinite incorporeal dimensions. But – because these incorporeal dimensions are supposed to be independent of God and uncreated – in order to guarantee the possibility of God creating an infinite world, infinite incorporeal dimensions must actually exist. Such a line of thought, I take it, is what lies behind Gassendi's move from the existence of incorporeal dimensions to their infinitude.

43 This is a main theme of Osler, *Divine Will and the Mechanical Philosophy*.

It is worth noting that the actual infinity of space does no work within Gassendi's physics. His atomism requires space to be distinct from body, but it does not require space to be infinite – only to be at least as big as the created world. He accepts the infinitude of space because he thinks his argument for absolute space implies it, but the infinitude of space has no further consequences.

That space is uncreated, and indeed is not the kind of thing that *could* be created, is doing a great deal of work here. However, as we shall see later, it is difficult to see how Gassendi can think that this is acceptable given that God is the author of all things. For he also holds that space is distinct from God, as the claim that absolute space must exist for God to exercise his creative power suggests. Gassendi's argument for a space like Patrizi's leads him to an ontology radically divergent from Patrizi's emanationism.

The Interstitial Vacuum

Gassendi's claim that an extra-mundane void exists was somewhat controversial, but there were ample historical precedents for it. Some of the figures discussed earlier, most notably Patrizi, argued for the existence of an interstitial or interparticulate void as well. Granting the possibility or even actuality of an extra-mundane void makes it easier to accept the existence of an interstitial void. However, because motion through the created world is evident, and because Aristotle's arguments that motion through a void is impossible remained on the table, the actuality of a void within creation remained in dispute. To rebut the Aristotelian objections, Gassendi's main tactic is to use Epicurean arguments that purport to show that motion is *only* possible if there is a void.

Now if this line of argument works at all, it should be entirely conclusive. Nevertheless, Gassendi also emphasizes the usefulness of the interparticulate void in explaining rarefaction and condensation, apparently ignoring the alternate explanations of these phenomena put forth by authors such as Descartes.

The argument that motion requires a void is famous and – in its most basic form – famously simple-minded, as Epicurus' critics pointed out. The basic premise is that motion and bodies are evident in sense perception. (Although the evidence of motion had been doubted, the premise is unproblematic in this context because the Epicureans are arguing not against skeptics but against other positive theories of matter.) Now, a guiding principle of Epicurean canonic is that the nonapparent must be inferred from the evident, and Gassendi's argument *If there is motion, then there is a void; but there is motion; hence there is a void* is such an inference (1.192b). Expanding on the first premise, he argues that if there were no void, there would be no empty place for bodies to move into and hence that motion would be impossible.

As is usual for inferences from signs, most of the work is being done by the bridge premise, in this case the claim that without empty space for bodies to move into they could not move at all. Lying behind this is the thesis that bodies are by nature impenetrable, so that no two bodies can be in the same place at the same time. Hence a body cannot move into a region of space already occupied by another body, so if motion is possible, there must be void space within the world. The obvious worry here is how Gassendi thinks he can rule out the Cartesian account of motion in a plenum or, if the argument was developed before Gassendi encountered Cartesian plenism, how he could rule out *Aristotle's* account of motion in a plenum.[44]

The plenist claims that motion does not require a void because everything is moving in vortices: When a body *a* moves into the place previously occupied by *b*, *b* simultaneously moves into the place occupied by *c*, *c* into the place of *a*, and so on. Aristotle, for instance, argued that

> not even movement in respect of place involves a void; for bodies may simultaneously make room for one another, though there is no interval separate and apart from the bodies that are in movement. And this is plain even in the rotation of continuous things, as in that of liquids.[45]

Lucretius, whom Gassendi discusses in the relevant sections of both the *Syntagma* (1.193a ff) and *Animadversiones* (*Animadversiones* 1.171 ff), replied to Aristotle using the example of fish swimming through a relative plenum, water. Even though there is no space between fish and water, the plenist might say, the fish can move by changing places with successive portions of the water that flows around them as they move forward. However, Lucretius argues against this intuitive characterization of the motion of fish:

> It is said that water makes way for scaly shiners and opens paths for them because fish leave behind them places where the waves can flow together. In the same

44 Bloch, *La philosophie de Gassendi*, 496 ff, dates the original composition of this section of the *Syntagma* to around 1637. However, it did undergo some later revision, so it is plausible that Gassendi did not think the argument needed to be revised in light of what he had learned about Cartesian plenism.

45 Aristotle, *Physics*, 4.7, 214a29–32.

way, other things can also move reciprocally and change places even though the whole world is full. But this reasoning is entirely false. For where could the fish move, in the end, if water did not leave space? Moreover, how could waves make way if fish could not move? Therefore, we must either deprive bodies of motion entirely, or say that void is mixed into the world and, because of it, bodies singly make their start toward motion.[46]

At *Animadversiones* 1.172, Gassendi reads Lucretius as pointing to the difficulty of explaining "by what reason motion could begin" in a plenum. The fish cannot start to move unless the water has already moved and left a vacant space, but the water cannot begin to move unless the fish has already moved, leaving an empty space for it to flow into. Thus, Gassendi might say, even if we grant that there is no problem understanding how Cartesian motion could continue once vortices were already established, it could never get started in the first place.

Unfortunately, it is not difficult for the plenist to respond. As a first pass, she could answer that there simply *was* no first motion and that the world has always been in motion. Gassendi can rule this answer out on the grounds that it is an article of faith that the world had a beginning in time. However, the plenist could easily amend her answer and say that although the world had a beginning in time, it was created already moving, so that there need be no first motion. It is hard to see how Gassendi could object to this second response. For we shall see that he himself is committed to the view that atoms were created endowed with *vis motrix* and hence most likely already in motion.

Another response that could be offered on Gassendi's behalf would emphasize that we experience not only motion but also motion stopping and starting.[47] The observed fact of beginnings and ends of motion could thus help us get around the worry that the world might have been created already in motion, so there is no need for a beginning of motion. However, the plenist again has a ready response. She could simply deny that genuine, micro-level rest need underlie the apparent macro-level rest we observe. Here again Gassendi's atomism seems to block any possible response because he himself denies that when composite bodies are

[46] Lucretius, *De rerum natura*, 1.372–80.

[47] This is suggested by Philodemus in *De Signis*. He claims that the premise *if there is motion then there is a void* is a generalization from observed instances of motion, where the space into which solid bodies move is always relatively empty before they move into it. Philodemus, *Philodemus: On Methods of Inference* 111 (Section 37). See also Asmis, *Epicurus' Scientific Method*, 209.

at rest their component atoms must also be at rest. I am unable to find in Gassendi a plausible answer to the plenist, so the argument that motion requires a void will have to remain notorious for the time being.

However, it is worth noting that, by the time Gassendi gives this argument for the existence of a void, he already takes himself to have defeated the chief motivation for plenism, namely the alleged impossibility of void space. Gassendi's atomism is motivated by its explanatory value. The argument that motion is impossible without a void is employed after the fact for justificatory purposes and, despite its allegedly demonstrative status, was never intended to bear anything like the full weight of establishing atomism.

The Coacervate Void

The extra-mundane void plays a crucial rhetorical role for Gassendi, allowing him to present readers with familiar, theologically acceptable arguments that void space was in some sense possible. Such arguments have the effect of softening the reader up in preparation for the more controversial argument that there is an interparticulate void. In contrast, the question of whether a macroscopic void can be artificially created is an empirical question, in principle entirely separate from Epicurean atomism. Proof that a coacervate void existed would confirm Gassendi's account of the nature of space, but his account of space does not commit him to the view that any coacervate void can actually be produced. Indeed, in early drafts of the *Syntagma*, Gassendi does not discuss the question of the coacervate void at all.[48] However, his interest in the topic must have been piqued by Evangelista Torricelli's 1643 barometer experiments and the controversy that developed in France as a result of their dissemination and Pascal's 1647 *Expériences nouvelles touchant le vide* and *Récit de la grande expérience à l'équilibre des liqueurs.*[49]

The basic form of the experiment is simple, requiring only a four-foot-long tube, hermetically sealed at one end and filled with mercury, and a bowl containing a layer of mercury under a layer of water. With your finger over the open end of the tube, turn the tube over and place the open end in the bowl's mercury layer. When you take your finger away, the mercury will drop, leaving an apparently empty space at the top of the tube. As the

48 Brundell, *Pierre Gassendi*, 62–3.

49 At 1.204 ff, Gassendi recounts his knowledge of Torricelli's and Pascal's experiments.

tube is raised higher in the bowl, the space grows progressively larger – until the mouth of the tube is raised above the bowl's mercury layer and into the water layer. At this point all the mercury in the tube will flow out and the tube will fill completely with water. The fact that the water fills the tube completely after the mercury flows out is supposed, on some accounts, to show that the apparently empty space at the top really was empty.

Differing conclusions were drawn from this experiment and its successors. Although a great deal of secondary literature has already been devoted to the differing interpretations of the barometer experiments, a brief summary of the positions on offer will help us understand Gassendi's response.[50] There were two main issues, as Gassendi notes: whether the space in the tube above the mercury was really a void and what supported the column of mercury (1.205a). Three answers were on offer: an absolute *horror vacui*, which implied that the space above the mercury was not actually void; a limited *horror vacui* of the sort Campanella propounded, which could be partially overcome by external forces;[51] or the weight or pressure of the air. The last two options left it open whether the space at the top of the tube was void or not, but Pascal was rather in the minority in thinking that the space was genuinely void.

Given the controversy surrounding interpretation of the results, Pascal proposed a further experiment that he thought would convince his audience that what suspended the column must be the weight of the air and not an absolute or relative vacuum. In 1648 – inspired by Torricelli's suggestion that the weight of the air, which is greater at sea level than at altitude, caused the column to remain suspended[52] – Pascal had his brother-in-law perform the original experiment again on top of a mountain, the Puy-de-Dôme.[53] The *Récit de la grand éxperience à l'équilibre des liqueurs* was published soon after Mersenne's death on September 1. Pascal's claim – which turned out, of course, to be true – was that a shorter column of mercury would be suspended at higher altitude. Because the weight of the air varies with altitude but the alleged *horror vacui* (whether absolute or relative) does not, this shows that the weight of the

[50] See, for example, Dear, *Discipline and Experience*; Fouke, "Pascal's Physics"; Garber, *Descartes' Metaphysical Physics*; Grant, *Much Ado About Nothing*; Mazaurac, *Gassendi, Pascal et le querelle du vide*.

[51] Campanella, *De sensu rerum*, 35 ff.

[52] Fouke, "Pascal's Physics," 86.

[53] Descartes also has some claim to the idea. See Pascal, *Oeuvres complètes*, 2.655 ff (editors' remarks) and Rodis-Lewis, *Descartes: His Life and Thought*, 179.

air was the cause.[54] Gassendi repeated this experiment in Toulon with the same result: The height of the mercury column varies inversely with altitude.[55]

Although Pascal claimed that "all [the philosophers] hold as a maxim that nature abhors the void,"[56] in fact few people argued against him on the grounds of a natural abhorrence of the void. The notable exception here is the Jesuit Étienne Noël, with whom Pascal had corresponded and whose objections were well-known. In a letter to Pascal, Noël objected that the apparently void space above the mercury "is a body because it has the actions of a body," one such action being that "it transmits light with refraction and reflection."[57] He found conceptual difficulties in the very possibility of a void, writing that Pascal's claim of a void in nature "is not only repugnant to common sense, but furthermore manifestly contradicts it: it implies that the void is a space and that it is not [a space]". For on Noël's view, "all space is necessarily body" because any space is composed of parts beyond parts, with a certain length, depth, breadth, and shape.[58]

In reply to Noël's worries, Pascal suggested that he should simply conceive of space differently – as something that possesses three dimensions, is immobile, is penetrable by bodies, and occupies a midpoint between matter and nothingness.[59] He added in a letter concerning Noël's objections that space (like time) is neither a substance nor an accident, "as they are commonly understood."[60] This conception of space is obviously similar to Gassendi's, and, indeed, Bloch suggests that Pascal's conception of space may have been inspired by the *Animadversiones*, a manuscript of which was then circulating around the Mersenne circle.[61] But because Gassendi's conception of space was by no means unique, it is difficult to know for sure.

Descartes, like Noël, granted the causal efficacy of the weight of the air but denied that the space above the mercury in the tube was a void.[62] Indeed, given his thesis that the essence of body is extension, he had to

54 Pascal, *Oeuvres complètes*, 2.679–80.

55 Rochot, "Comment Gassendi interprétait," 55–6.

56 Pascal, *Oeuvres complètes*, 2.678.

57 Noël in Pascal, *Oeuvres complètes*, 2.513.

58 Ibid., 2.517.

59 Pascal, *Oeuvres complètes*, 2.526.

60 Ibid., 2.565.

61 Bloch, *La philosophie de Gassendi*, 196–8.

62 For Gassendi's account of Descartes's view, see 2.482, although this is very brief. Compare Garber, *Descartes' Metaphysical Physics*, 136 ff.

deny that there could be a void above the mercury. Gassendi also denied that the space above the tube was a vacuum. He adopted a portion of Noël's argument, claiming that the space above the tube could not be a vacuum because light and gravity – both of which are, on his view, corpuscular – passed through it (1.206b). Of course, he did not sympathize with Noël's worries about the conceptual possibility of a void. Nor did he dignify them with a response beyond that implicit in the accounts of space provided in the *Animadversiones* and *Syntagma*.[63] Rather, he took the experiment to provide further evidence for the possibility of a vacuum, for two reasons. First, he thought the barometer experiments showed that at least relatively void spaces could be created in nature: The space above the tube was on his view devoid of air, if not of light, gravity, and magnetism. Gassendi seems sure that the glass tube does not admit air and that, because mercury is heavier than air, there was no air mixed with the mercury to begin with (1.205a–b). Second, Gassendi offered an explanation, in terms of atoms and the void, of the difference in the height of the column at various altitudes, and he took the plausibility of such an explanation to count in favor of atomism.

Gassendi's explanation relies on distinguishing air pressure from the weight of the air. Air pressure, on his view, is a product of the weight of the air: Air is pressed down by the weight of the air above it, and the corpuscles of air resist being pressed together so that the intervening void space is diminished (1.198a–b). Thus, Gassendi interprets the results of the barometer experiment as follows. The weight of the layers of air higher in the atmosphere compresses the layers of air nearer the earth, and the lowest layers of air press against the mercury in the bowl, preventing it from rising as much as would be necessary for the mercury in the tube to descend fully. When the barometer is at altitude, the relevant layers of air are less compressed so that there is less pressure on the mercury in the bowl (1.213b). Gassendi argues that the different height of the column at different altitudes shows that the relevant factor must be air pressure and not the weight of the air. Given his account of gravity, both the air and the mercury in the tube should be less affected by gravity on a mountaintop; that is, both the weight of the air and the weight of the mercury change so that the ratio remains the same (1.208b). Thus, air pressure and not weight must be the cause.

[63] Gassendi's discussion, inserted into the *Syntagma* as a chapter, was written in 1647 or 1648 and circulated as a letter although it was not included in the *Animadversiones*. Rochot, "Comment Gassendi interprétait," 54.

Because Gassendi did not claim that the space above the mercury was a genuine void, his interpretation of the barometer experiments is in some sense orthogonal to the issue of the void. His discussion of the coacervate void occupies two chapters of the *Syntagma* – one devoted to the barometer experiments and another shorter one to the instantaneous voids produced by machines, most notably the machines from Hero of Alexandria's *Pneumatics* (1.199b ff). It was obviously important for Gassendi to integrate the barometer experiments into his theory, given their prominence in his intellectual circles. However, the chapters on the coacervate void are rather uneasily sandwiched into the middle of a generally much more abstract book on space and time. Here, as elsewhere, Gassendi is concerned to show that particular well-known phenomena accord well with his atomism, without ever making the stronger claim that such phenomena constitute a proof of it. The only argument for atoms and the void that is ever claimed to provide certainty remains the troublesome argument that motion requires a void.

Space and God

We have already seen three of the key features Gassendi ascribes to space: It is three-dimensional, incorporeal, and immeasurable or infinite in all direction. Because Gassendi's discussion makes clear that all regions of space are the same and that space has no center that could ground a natural direction of motion, we can add that space is homogenous, although Gassendi does not use the term. Space is also immobile, as the thought experiment about God annihilating and recreating all the bodies beneath the lunar sphere shows (1.183a–b). If spatial dimensions are distinct from the dimensions of bodies, they need not move when bodies move. Rather, motion is a matter of going from one region of space to another. Indeed, Gassendi *defines* motion as transfer from one place or region of space to another so that we cannot even make sense of space moving or being moved.

Space's dimensionality implies its mathematical divisibility, which for Gassendi amounts to purely conceptual divisibility, licensing no inferences about the physical world. The mere fact that we can conceive of a limited region of space shows that space is divisible in this sense. Moreover, Gassendi argues, we can conceive of space as containing any geometrical figure we like, and because geometrical figures are composed of infinitely divisible lines, it follows that space must be infinitely mathematically and conceptually divisible (1.246a).

However, no division of space can actually be performed because space is by nature one and uncuttable. To see why this should be so, consider what physical divisibility really amounts to on Gassendi's account. To actually divide something (rather than merely imagine dividing it) is to cut and separate its parts. But because space is incorporeal and penetrable, it cannot be cut. And because it is immobile, its parts cannot be separated. Nor could one make a real division in space by removing a part of it, leaving the rest unchanged, for space is by nature indestructible: That which is uncreated cannot be destroyed. Hence, no real divisions can be made in space.

Penetrability is supposed to follow from the incorporeality of space. Because spatial dimensions are incorporeal, they have "no repugnance to bodies either penetrating them or resting within them" (1.183a). Impenetrability (or, as Gassendi takes to be synonymous, resistance or solidity) is criterial of bodies. This claim is crucial for it allows Gassendi to avoid the conclusion that space is itself a body. At the same time, however, Gassendi's attempt to render unobjectionable the claim that space is uncreated and independent from God requires that the penetrability of space is not a positive property or power, but merely a privation.

Because penetrability precludes acting or being acted on by contact, the incorporeality of space also implies its inertness.[64] Space, Gassendi says, "cannot act or be acted on in any way, but has only a negative quality [*repugnantia*] by which it allows other things to pass through or occupy it" (1.183b). Indeed, all the properties of space save three-dimensionality (and its corollary, mathematical divisibility) should be understood as privations, for this helps us see how Gassendi can maintain that space is not a positive being.

Gassendi makes clear that a crucial mark of a positive being is having the power to act so that inert space is nothing positive:

> [W]hen we say that there is an incorporeal interval and incorporeal dimensions, it is so clear it really need not be said that this is another kind of incorporeal than that which is a species of substance and fits Almighty God and the Intelligences and the human mind. Of course, in this latter case the word "incorporeal" does not mean only being destitute of body and the negation of corporeal dimensions but also refers to a true and genuine substance and a true and genuine nature whose faculties and actions fit it. (1.183b)

[64] As we saw in Chapter 2, in the beginning sections of the *Syntagma*, only God can act without contact. This, of course, raises problems concerning the action of the incorporeal soul, which I discuss in Chapter 10.

The capacity to be acted upon is another mark of a positive being, assuming that to be a substance is to be a positive being (1.184b), and space lacks this mark as well.

I have left for last the most problematic properties Gassendi ascribes to space. Space is uncreated, *improductum*, and eternal or – as the discussion of time will make clear – sempiternal. Since God is the author of all things, it causes problems to hold that space is uncreated.[65] The *Syntagma* suggests two different ways of dealing with these problems. One involves claiming that space is not really a thing or, if you prefer, that God is the author of all *substances* but not all *things*. (Which you prefer depends on whether you find it acceptable to say there are things that are "nothing positive.") The second consists of claiming that space depends on God without being created by him – as, for instance, modes depend on substances without being created by them – so that space is not wholly distinct from God. This is, famously, Newton's account of the relation between absolute space and God. These two suggestions work rather at cross purposes. Someone who takes space to be an attribute of God does not really need a separate argument that space is not the sort of thing that needs to be created. Nevertheless, Gassendi suggests both answers.

First, let us look at Gassendi's suggestion that space is not a genuine thing and, hence, need not be created. Several of his claims point in this direction: the note that space is nothing positive; the rather misleading comparison between his own view and the scholastic view of imaginary space; and the comparison between the reality of space and the reality of essences. The following passage, for instance, contains all three suggestions:

> [W]e must eliminate any scruples that might arise from the following. It might be inferred that space, as it has been defined here, is not produced by God and is independent of God. And because it has been said that space is something, it appears to follow that God would not be the author of all things. But it is clear that by the words "space" and "spatial dimension" nothing is understood except that space commonly called imaginary that the majority of the holy doctors admit exists beyond the world. And they do not call it imaginary merely because it depends on the imagination like a chimera, but because we imagine its dimensions as being like the corporeal dimensions that fall under the senses. Nor does

[65] See, for instance, Berkeley's remark about "that dangerous *dilemma*, to which several who have employed their thoughts on this subject imagine themselves reduced, to wit, of thinking either that real space is God, or else that there is something beside God which is eternal, uncreated, infinite, indivisible, immutable" (Berkeley, *Principles of Human Knowledge*, §1.117).

it cause them problems that this space is said not to be produced by God and to be independent of God. For it is nothing positive, that is, neither a substance nor an accident, which two words together include everything produced by God. In fact, it appears that this is far more tolerable than something else the Doctors commonly admit, namely that there are essences of things that are eternal and independent of God and not produced by God. (1.183b–4a)

By comparing his notion of space to imaginary space, Gassendi is careful to note, he does not mean that space is purely conceptual. He calls space "imaginary" simply because we imagine it by analogy with corporeal dimensions. This explains how we can cognize space despite having no direct perceptual contact with it. But it is also a rhetorical trick, making Gassendi's view of space sound less radical than it actually is. Those who held that imaginary space existed outside the world did not hold that such space was a real thing cognized through imagination rather than sense or reason. Scholastic imaginary space did not, for instance, ground the possibility of a natural vacuum. I suppose certain readers may have found the suggestion about imaginary space reassuring because it seems to tie Gassendi's view in with the tradition, but it does not really help solve the problem.

Next, consider the suggestion that when it is said that God is the author of all things, this means that God is the author of all substances and accidents, thus making no mention of space. This suggestion relies on the claim that space is nothing positive, which I find hard to understand given Gassendi's insistence that it is a "true thing" and a "real being" (1.179a). Space *does* meet the first criterion of substantiality, existing per se, so the *sui generis* ontological status of space does not explain why an omnipotent God need not have created it.

Gassendi's third suggestion is rhetorically helpful but, again, does not really solve our problem. He writes that allowing space to be uncreated is no more worrisome than the commonly sanctioned view that essences are not created:

[That space is *improductum*] is far more tolerable than something else the Doctors commonly admit . . . namely, that there are essences of things that are eternal and independent of God and not produced by God, essences that are the principles of substances and accidents. For suppose it may be asserted that the essence of man never began and will never end. Because that essence consists in the fact that man is a rational animal, this [fact] will be necessary, so that it does not depend on God and cannot be made otherwise by any power. Then why could we not say the same thing about space, since it is not the sort of thing that is capable of being produced – unlike man, in whom the essence is the principle? (1.184a)

Gassendi offers the following argument for why it is worse to hold that there are uncreated essences than that there is uncreated space. Essences, as the principles of substances, constrain which substances can be created and in what way. Thus, to allow that essences are not created by God and are independent of him is to allow that there are external constraints on what God could create. But putting external constraints on God is far more worrying than holding that there is some ontologically marginal thing that he did not create. Gassendi's space does not constrain God's actions but rather is what must exist in order for God's creative actions to be possible.

There are two problems with this argument. First, as we saw in Chapter 2, Gassendi makes quite clear in the *Disquisitio* that he thinks it is entirely unacceptable to hold that there are essences independent of God. Indeed, he supposes that Descartes's view of the eternal truths requires them to be independent of God, at least once they are created, and takes this to count as a clear refutation of the Cartesian view of created eternal truths. Hence, it is no great consolation that the situation with regard to space is no worse. Second, it is simply not true that the view that there are uncreated essences independent of God was generally considered acceptable, and Gassendi must have been aware of this. He is defending a controversial position about space by appeal to another controversial position about essences. It may be that all Gassendi intends to establish by the comparison of the reality of space and the reality of essences is that his account of space is not immediately ruled out on theological grounds; certainly, the argument succeeds in establishing no more than that.

One common way to explain how essences can be uncreated is to suppose that they are dependent on God. Gassendi's second suggestion as to how to accommodate the uncreated nature of space, that space is "as it were the immensity of God," is a version of this:

> [B]ecause it follows from the perfection of the divine essence that it is eternal and immense, from this all time and all space are connoted, without which neither eternity nor immensity can be understood. And therefore God is indeed *in se* most highly and infinitely perfect, but nevertheless is also in every time, and in every place. And, when it is asked, where was God, before he founded the world, indeed it cannot be denied that he was *in se*, but at the same time it must be conceded that he was everywhere, or in every place, that is, not only in that space in which the world would be, but also in an infinity of other places. And that God is in space is considered an extrinsic denomination with respect to his essence, but not with respect to his immensity, the concept of which necessarily involves

the concept of space. For otherwise, that God is said to be his own place would clearly be metaphorical. (1.191a)

Tom Lennon describes how later Gassendist philosophers developed accounts of space along the lines of the two suggestions found in Gassendi.[66] One group – Gilles de Launay and Marin Cureau de La Chambre – adopted the suggestion that space is a feature of God and tried to clarify it. Another group, or at least Bernier, minimized and finally eliminated any real ontological status for space. This second move creates metaphysical problems for atomism. The move to make space a feature of God is more attractive, but, as is clear from the later debates about Newton's view of space, it remains controversial from a variety of Christian viewpoints.

Time

Like space, time exists independently of the things in it: There was a time when no world yet existed. It is clear why atomism requires absolute space but rather less so why it should require absolute time. Gassendi developed his account of time much later than his account of space, more or less just taking over the properties of space and applying them to time.[67] As early as 1631 he planned to treat time "differently from Aristotle" (3.102), but this is perfectly consistent with the possibility that he planned to treat it in the Epicurean fashion, as an accident.

Gassendi tells us that "on account of the remarkable parallelism that there is between place and time, it is necessary to treat the time or duration of things in the same manner" (1.133a). However, he says less than one might like about what that parallelism consists in. He suggests that one reason time is taken to be so mysterious is that it (like space) has traditionally been forced into the misleading schema of substance and accident. He notes that Aristotle discussed space and time in adjacent books and that Plato grouped them together (1.224a). And he compares space and time explicitly in the following terms:

[Time] seems to be something incorporeal and *per se* like void, that is, independent of the existence of any other thing . . . just as there is an incorporeal space that although called imaginary is nevertheless the same as that [space] in which the nature of place consists . . . in the same way there is a certain incorporeal duration

[66] Lennon, *The Battle of the Gods and the Giants*, 119.

[67] For an account of the development of the theory of time, see Bloch, *La philosophie de Gassendi*, 178–89.

independent of bodies, which although called imaginary is nevertheless the same as that [duration] which constitutes the measure of time. (1.220a–b)

Thus, he describes time as "an incorporeal flowing extension in which it is possible to designate the past, present, and future so that every object may have its time" (1.220b). A few pages on, he makes the striking claim that time has "successive . . . dimensions," corresponding to the motion of bodies, while space has simultaneous ones (1.244b).

What are the antecedents of Gassendi's view of absolute time or of space and time as parallel? It is not, of course, the Aristotelian view. Nor is it Epicurean: While space is for Epicurus an "intangible substance,"[68] time is an "accident of accidents,"[69] as Gassendi himself is careful to note (1.221b). Gassendi writes that the Stoic view of time as "the sort of incorporeal which is understood to exist *per se*," so that time can exist even if things in time do not, is much preferable to the Epicurean view (1.221b). However, he does not make terribly clear *why* it is preferable. Even though he says that if we conceive of God as destroying every created thing, time itself will still flow on so that there can be time without change, it is not immediately clear why the reader is supposed to grant this. The claim is not backed up by the kind of conceptual work evident in the case of space, and indeed it seems that the reader is already supposed to have accepted that space and time are ontologically parallel by this point. However, as far as I can tell, the view of time as independent of change and the things in it is neither required by anything else in Gassendi's natural philosophy nor argued for on independent grounds.

The best I can do for explaining the status of time is this. Given his deconstruction of the categories, Gassendi holds that Aristotelian ontology only allows substances, accidents, and mental entities or ways of conceiving. Time clearly cannot be a substance, for the same reason that space could not be a substance, namely, that things are not predicable of it. However, Gassendi cannot allow time to be an accident of bodies either. For one thing, this would create different, inconsistent times, flowing at different speeds – and if God had created many worlds, we would have many unrelated times (1.223b). This pushes Gassendi to place time outside the Aristotelian ontological scheme altogether. At this point it

68 Sextus, *Adversus Mathematicos* 3.211 (Section 10.2). Compare Diogenes, *Lives of Eminent Philosophers*, 40.1.

69 Sextus, *Adversus Mathematicos*, 3.319 (Section 10.219). Compare Lucretius, *De rerum natura*, lines 4.459–61: "time is not *per se*, but our sense of what happened in the past, what is happening now, and what is going to follow it [arises] from things themselves."

might seem entirely natural to group space and time together. After all, Gassendi points out, both are limitless, both are required for the existence of anything else, and both have dimensionality (1.224b).

I am aware of only one suggestion in the literature for why Gassendi maintains the parallelism of space and time. Barry Brundell argues that

> Gassendi could well appreciate their analogous character through his understanding of the new mechanics which represented space and time as mathematical co-ordinates ... Gassendi would seem to have foreshadowed philosophically the mathematico-physical innovation of Galileo when he described the "parallelism" of space and time in ... 1637, on the eve of the publication of the *Discorsi*.[70]

This strikes me as rather optimistic, for several reasons. First, even Gassendi's overtly Galilean *De motu* never represents either space or time in coordinate form. Second, it seems improbable that Gassendi would let such a mathematical representation determine his metaphysics, given his general insistence on separating mathematics from physics. And third, it is hard to make the coordinate view of time fit in well with Gassendi's language of the "flow" of time and his insistence that only the fleeting present exists.

Let us put aside the puzzle of why space and time are supposed to be parallel and turn to what Gassendi *does* tell us about time. He reverses the Aristotelian definition of time, arguing that time does not depend on motion but "is something only revealed by motion, as something measured is revealed by what measures it" (1.225a). What constitutes time is not measured motion but mere successiveness. In responding to the objection "that time is nothing, because, although it is said to consist of the past, the present, and the future, the past no longer exists, the future does not yet exist and the present is entirely evanescent" (1.222a), Gassendi replies that the people who make this objection

> consider heterogeneous things as if they were homogeneous, that is, consider successive things as if they were permanent, when they are completely different in kind.... They demand from successive things what is not in their nature, and if it were in their nature they would not be successive. For make their parts stand still, make them not flow, make them remain, then they will be not successive, but permanent. (1.222a)

Besides its flow and its successive dimensions, a few other features of time are worth noting. Time, like space, is entirely homogenous and is

[70] Brundell, *Pierre Gassendi*, 63. Sorell, "Seventeenth-Century Materialism," 248, makes a similar suggestion.

unaffected by and indifferent to whatever goes on in it (1.224b). And it is infinite, in the usual sense of being unlimited. In a letter to Sorbière of 1644, Gassendi writes that "both [space and time] are infinite *ex se*, the former according to the dimensions length, breadth and depth, the latter according to the succession prior and posterior" (6.179a). Just as space has infinite spatial extension, there is an infinite succession of time both before the beginning of the world and after its destruction (1.220b). Thus, time, like space, is uncreated.

However, even though Gassendi suggests at various points that space is also *independent* of God, he makes no such suggestion with regard to time. The distinction between God's endless duration and an endless duration taken abstractly is purely notional, and Gassendi pretty clearly opts for the solution that time is an attribute of God. This leads to his equation of eternity with sempiternity: "Eternity cannot be understood as anything else than perpetual Duration . . . inasmuch as it lacks beginning and end" (1.225b). "Gassendi suggests that the error of distinguishing eternity from sempiternity derives from Boethius's reading of the Timaeus. (He blames many philosophical mistakes, such as belief in a world soul, on the influence of the Timaeus.)" For, Gassendi argues, just as everything either exists in some place or does not exist at all, so everything either endures through some time or does not exist at all (1.224b).

The World

Gassendi draws a clear distinction between the status of space and time and the status of the created world, the collection of atoms making up all corporeal bodies. He also distinguishes their extents, as one would expect given the connection between the uncreated nature of space and time and their infinite extent. For Gassendi, the world is a finite island in an infinite sea of empty space and time. Here he diverges from the Epicurean position. Neither the plurality of worlds nor an infinitely long world were acceptable within Christianity, and hence looking at this case helps us understand how Gassendi thinks about issues significant for both religion and natural philosophy.

The Epicureans held that space is infinite and there are infinite atoms, but that each world is made up of a limited number of atoms, so that there is an infinity of worlds. Gassendi's view is more similar to the Stoic view, revived by Patrizi in his *Nova de universis philosophia*, that there is a finite world existing in infinite space. Gassendi endorses the one-world view on the grounds that it is "in the greatest consonance with the principles of sacred faith and religion. It is clear that the sacred writings have no mention of a multitude [of worlds]; and whatever they relate concerning the origin of things, providence, the condition of men, happiness, and other things, they indicate a singular world" (1.141a). Gassendi holds that the one-world view cannot be demonstrated "because . . . God could and still can found innumerable other" worlds (1.141b). Nevertheless, he insists that "it is beyond all reason that there are in fact many worlds, both because God plainly wanted nothing to become known except concerning this one, and because the opposite reasons are either completely frivolous or do not have much probability" (1.141b). Our ignorance of other possible worlds makes it probable that no other worlds exist because, Gassendi assumes, God would have no reason for concealing them from us if they did exist. It is hardly surprising that Gassendi is concerned to defend the one-world view, considering that one of the things Bruno had been condemned for was championing the plurality of worlds.

Orthodoxy similarly requires holding that the world has a beginning in time. As Gassendi notes, Genesis begins "In the beginning God created heaven and earth" (1.163a). Hence, the view of Epicurus that the world had a beginning in time was correct, even though Epicurus held this view for the wrong reason. Similarly, Gassendi says, Epicurus' claim that the world will come to an end is correct although his suggestion that this will come about through natural processes is wrong (1.171b).

Gassendi makes clear that reason cannot demonstrate the beginning or end of the world because, according to reason, it is perfectly possible that the world is eternal (to Gaffarel, 1629; 6.12b).[71] However, in the *Syntagma,* he is more concerned to emphasize that probable reasoning favors the beginning of the world. The same sort of empirical evidence that he takes to show with the greatest probability that the world has a designer also thereby shows that the world has a beginning. For

> if this order [of nature] was established by some cause, then it is not eternal. For establishing is an action that happens at some time . . . an eternal thing does not ever become and is not established at some time. Rather, that which is all at once must always exist, and is considered as a pure perseverance through the whole series of times. (1.163a)

No similar argument is given for the end of the world; Gassendi seems to endorse the view that we know the end of the world only through

[71] This early remark is not necessarily conclusive, however. We know that in the *Exercitationes* Gassendi held that the existence of God was indemonstrable but offered alleged demonstrations of it by the early 1640s, in the Fifth Objections, and later in the *Syntagma.*

revelation. Here he is taking part in an old dispute about whether there was rational evidence of the beginning of the world and if so whether that evidence was demonstrative.[72] In Gassendi's tradition, it was generally agreed that there was no demonstrative proof of the beginning of the world. However, he wants to make clear that there is some evidence beyond revelation and Church dogma that the world had a beginning in time. Probability supports and accords with what faith demands, and the harmony between faith and probability serves to reinforce the desirability of probable knowledge of the world.

[72] Sorabji, *Time, Creation and the Continuum*, 193–202.

6

Atoms and Causes

Early modern atomists and corpuscularians adopted their view for a number of different and sometimes inconsistent reasons and understood their atoms in a great variety of ways. Some, but by no means all or even a clear preponderance, thought that atoms moved through a void. Others – like Basso, Sennert, and Descartes – held that there were atoms or *minima naturalia* or stable small corpuscles, but that the world was a plenum. In some versions of corpuscularianism, atoms were all composed of the same, homogeneous matter, differing only in size and shape. In others, particles or atoms were understood as materially variegated so that there were particles or atoms of earth, air, fire, and water, for instance, or salt, sulfur, and mercury. Such views were often closely related to the *minima naturalia* tradition that can be traced back to Aristotle's remark that

> since every finite body is exhausted by the repeated subtraction of a finite body, it is evident that everything cannot subsist in everything else. For let flesh be extracted from water and again more flesh be produced from the remainder by repeating the process of separation; then, even though the quantity separated out will continually decrease, still it will not fall below a certain magnitude.[1]

Substances, that is, can be broken down into finite parts that are the smallest possible unit of that substance. Thus, there are *minima* of each substance, animal, vegetable, or mineral. Some early modern writers – like

[1] Aristotle, *Physics*, 1.4, 187b256 ff. Compare *De generatione et corruptione*, 1.2: "[A] body is in fact divided into separable magnitudes which are smaller at each division . . . the process of dividing a body part by part is not a breaking up which could continue *ad infinitum*" (316b27–31). This appears to be a summary of an atomist position that Aristotle goes on to reject, but the similarity to the *Physics* passage is striking.

Daniel Sennert – invoked Aristotle as the first minimist, thus adding an aura of legitimacy to their view.[2]

Minima are not, of course, atoms. Indeed, in *De generatione et corruptione* Aristotle argued that atomism could not explain how new substances were generated in chemical mixtures: Simply combining atoms cannot make genuinely homogeneous substances. To "the eyes of Linceus," who can see beyond normal human capacities, there will be no mixture but only brute association.[3] But later corpuscularians, including those who, like Sennert, had a notion of atoms, saw a way around this difficulty. They claimed that such mixtures were composed of homogeneous corpuscles or molecules, although those corpuscles themselves were internally heterogeneous. This solution requires a principle of unity for the internally heterogeneous molecules, and we shall see that Sennert offers substantial forms to provide such a principle.

In other varieties of atomism, atoms were materially homogeneous, differing only in size, shape, and motion or motive power. This, according to Gassendi, is the view of the ancient atomists Democritus, Epicurus, and Lucretius. It is also the view sketched in the opening sections of the *Syntagma* and the view supposedly demonstrated by the existence of motion. However, we shall see that by appealing to corpuscles – small aggregates of atoms with sufficient stability to figure as basic principles in explanations – Gassendi also manages to co-opt the qualitative explanatory patterns of theorists who subscribed to a non-Epicurean version of corpuscularianism.

Still other atomists did not even understand atoms as material beings, holding that they were mathematical points or indivisibles that composed the continuum: This is the view most common in the fourteenth century and famously associated in the early modern period with Galileo.[4] Gassendi is generally careful to distinguish between mathematical and physical atomism, holding that the former has no bearing on the latter and that physical atoms are infinitely mathematically divisible. Instead of responding to medieval criticisms of atomic indivisibles, he simply insists that they are irrelevant because we are now concerned with bodies and

[2] For instance, Sennert, *Hypomnemata Physica*, 91 (Section 3.1) cited Aristotle's *De generatione et corruptione* 8 as an antecedent. There is a great deal of secondary literature concerning *minima naturalia* theories. See, for instance, Emerton, *The Scientific Reinterpretation of Form*; Clericuzio, *Elements, Principles and Corpuscles*; or Murdoch, "The Medieval and Renaissance Tradition of *Minima Naturalia*."

[3] Aristotle, *De generatione et corruptione*, 1.10, 327b33–328a18.

[4] Murdoch, "Infinity and Continuity," 575 ff.

not mathematical entities (1.232a). Perhaps all the adherents of chemical atoms or *minima naturalia* made similar claims; at any rate, Sennert insists on the same point.[5]

Some atoms were efficacious in virtue of their souls, as Bruno's were.[6] Others, like Sennert's, were efficacious in virtue of their supervening form.[7] Still others, like Gassendi's, were efficacious in virtue of their motion or motive power – although Gassendi ascribes powers beyond the motive power of the component atoms to certain corpuscles. Similarly, a great diversity of arguments was advanced in favor of atomism and minimism.[8]

One might well wonder why all these various entities were known as "atoms," or whether the different forms of seventeenth-century corpuscularianism have anything in common beyond an appeal to indivisibles of either the mathematical or the physical kind. However, very many seventeenth-century atomists claimed the heritage of Democritus, Empedocles, and Epicurus for their view.[9] Democritean lineage was invoked even by those atomists whose atoms seem very different to us from Democritus's atoms; for instance, Sennert cites both the "atomism" of Democritus and the "atomism" of Aristotle's *De generatione et corruptione* as anticipations of his view.[10] Moreover, this initially puzzling terminological similarity allowed very different forms of atomism to influence each other. Gassendi moves easily from the insistence that atoms possess only size, shape, and motive power to a view on which they have all sorts of additional qualities in part because his sources call both sorts of things atoms and allow there to be compound atoms with powers additional to the powers of simples.

I shall begin with a brief account of some of the varieties of atomism on offer in the early seventeenth century. My intention is not to provide a

5 Sennert, *Hypomnemata Physica*, 91 (Section 3.1).

6 Gatti, "Giordano Bruno's Soul-Powered Atoms," 170–3.

7 In *De chymicorum*, Sennert claims, "natural things do not come about through the accidental aggregation of atoms... but by means of the direction of a higher form, which by the instrument of heat attracts, restrains, mixes and organizes everything as it works." Cited by Michael, "Sennert's Sea Change," 357. The same claim is made in Sennert, *Hypomnetata Physica*, 133.

8 A taxonomy of such arguments that aims at showing the minimal role played by *experimental* arguments is Meinel, "Early Seventeenth-Century Atomism."

9 Lüthy, "The Fourfold Democritus on the Stage of Early Modern Science." Lüthy describes various ways Democriteanism was understood and how textual confusion, including a number of spurious works, gave rise to so many different ways of understanding Democritean atomism.

10 Sennert, *Hypomnemata Physica*, 89 (Section 3.1).

complete history of atomism but merely to help us understand *Gassendi's* atomism, so I shall focus on two relatively well-known texts: Sebastian Basso's 1621 *Philosophiae naturalis adversus Aristotelem* and Daniel Sennert's 1636 *Hypomnemata physica*. Although I do not claim any precise connection between the views of these authors and Gassendi, it is helpful to see what other sorts of atomist or corpuscularian views were current. This will be particularly useful for understanding the relationship between Gassendi's Epicurean account of the phenomena of inanimate bodies and the minimist and often anti-reductionist accounts he offered for generation and various chemical phenomena.

I then move on to the character of Gassendi's atoms. Two properties of atoms, size and shape, are entirely unproblematic. A third property, which Gassendi variously refers to as *pondus*, *gravitas*, and *vis motrix*, requires somewhat more interpretive work. On one reading, Gassendi's atoms are always in motion with a constant speed and undergo mere change in direction as a result of collision. On another, weaker reading, it is merely the motive power of individual atoms that is conserved, not their actual motion. This second reading may strike contemporary readers as more palatable. It certainly seems to fit in better with Gassendi's Galilean account of the motion of composite bodies. However, I argue that the former reading accords better with Gassendi's overarching theses about causation. Whatever else *vis motrix* is, it is that in virtue of which atoms are active, and thus that in virtue of which there is secondary causation in the created world. We have already seen Gassendi's argument that neither Aristotelian forms nor a neo-Platonic world soul can successfully account for the activity of the created world. Here we shall see why motive power can succeed where forms and souls failed.

I shall also spend some time on the accounts of conservation and concurrence that are relevant to Gassendi's claim that matter is active. For although Gassendi himself does not see conservation and concurrence as central issues for physics and does not appear to take seriously the possibility that concurrence is incompatible with secondary causation, occasionalism became prominent so soon after his death that it is worth looking at how his system avoids the occasionalistic implications often seen in Cartesianism.

Varieties of Atomism

In addition to the Democrito-Epicurean atomism that was to become central for mechanical philosophers, a number of other atomistic traditions were present in the sixteenth and seventeenth centuries. Two significant traditions were the *minima naturalia* tradition and the various forms of chemical atomism found in writers such as Severinus, de Clave, and van Helmont.[11] These traditions come together very clearly in the work of Daniel Sennert and Sebastian Basso, the atomists whose views I shall sketch for the purposes of comparison and contrast.

In his charming *Hypomnemata physica*, Sennert advances an atomist position that developed in the context of sixteenth- and seventeenth-century debates about mixtion theory.[12] Mixtion, within an Aristotelian framework, occupies a place intermediate between mere alteration (where a continuing substance changes qualities) and generation (where an entirely new substance comes into being). Thus, mixtion theory is used to explain how, for instance, a new metal is produced from compounding two others, or how animal bodies produce blood and bone from the food they ingest.

As one might imagine, there was a fair amount of disagreement about what mixtion was and even about whether certain stock cases were genuine mixtures. Thus, Sennert begins by noting that mixtion is not always understood in the same way.[13] In one sense, you have mixtion when two or more things are combined in such a way that although the parts preserve their individual nature and could be separated, nevertheless you have something that is called one body. An example is "a body composed from wine and water, which is not one by form but only by continuation."[14] He compares this to the case where a heap of millet and a heap of wheat are commingled: Such a case differs from the wine-water case, Sennert argues, only in that we ordinarily tend to call the millet-wheat mixture a mere aggregate and the water-wine one a unified thing. The more interesting sense of the term "mixtion" applies to "a body composed of many things so that from all of them is made a certain one thing, different in

[11] Severinus, *Idea Medicinae Philosophicae*; de Clave, *Nouvelle lumière philosophique des vrais principes et élémens de nature*; Van Helmont, *Ortus medicinae, id est, initia physicae inaudita.*

[12] It is worth noting that Sennert's views underwent significant change from the early textbook the *Epitome*, cited in the preceding chapter, to the 1619 *De chymicorum consensu ac dissensu* and the 1636 *Hypomnemata physica*. Indeed, the *Epitome* itself also went through several significantly different editions. See Michael, "Sennert's Sea Change"; Newman, "Experimental Corpuscular Theory"; Michael, "Daniel Sennert on Matter and Form."

[13] Sennert cites "*Exerc. 101 sect. 1*" (that is, section 101.1 of Julius Caesar Scaliger's 1447 *Exoticarum exercitationum liber quintus decimus de subtilitate ad Hieronymum Cardanum*), an attack on Girolamo Cardano's popular *De subtilitate rerum*, as an example of the debate.

[14] Sennert, *Hypomnemata Physica*, 118.

nature from all the components and possessing a different form than the form of the components."[15] A perfect mixture like this is "*unum formaliter*," whereas the water-wine mixture or an alloy of gold and silver is not "*unum formaliter*... but rather *unum per accidens*."[16] The qualities of an accidental unity derive from the forms and qualities of the components; the qualities of a formal unity derive from a new supervening form. New supervening forms derive from God, but (at least in the case of inanimate natural bodies) Sennert is agnostic about whether each token form was created as an individual, or whether God created species capable of replicating themselves.[17]

Sennert thinks that it is relatively uncontroversial that a new supervening form is taken on in perfect mixtion[18] and even ascribes belief in new supervening forms to Empedocles, Democritus, and Anaxagoras.[19] For him, two questions are controversial: whether the ingredients of a mixture lose their original form when they take on a new supervening form, and whether perfect mixture only takes place among elements or whether other bodies or "nobler elements" can also enter in mixture. Let us consider the first question first. Sennert cites Empedocles, Democritus, and Anaxagoras – along with Scaliger, Albertus Magnus, Peter Aureole, Avicenna, and Jean Fernel – as holding that "bodies that are mixed retain their forms in mixtion."[20] In contrast, the antiatomist view that miscibles lose their forms in mixture is Aristotelian, at least according to *De generatione et corruptione* 2.7, although Sennert notes that Aristotle's interpreters have "waged civil war all through the ages" on this point.[21]

Sennert puts forth his own view as a middle way between bare aggregation and the Aristotelian view. He does not devote any real attention to criticizing the bare aggregation view because he does not think anyone holds it, but we shall see when we get to Basso that the fundamental

[15] Ibid., 119.
[16] Ibid., 120.
[17] Ibid., 135.
[18] Ibid., 133.
[19] Ibid., 120.
[20] Ibid., 123.
[21] Ibid. For instance, he describes Averroes as holding that both the substantial form and the qualities of the elements remain in mixture, although "weakened and restrained"; Scotus as saying that both forms and qualities utterly disappear; and "many Latins" as saying that the forms utterly disappear but the qualities remain, although weakened. This last group includes Zabarella and Toletus, in 1.10 of his commentary on Aristotle's *De generatione et corruptione*. See Toletus, *Commentaria... in Octo Libros Aristotelis de Physica*, 298a ff.

problem for the bare aggregation view is explaining how aggregation can give rise to entirely new qualities. Sennert attacks the Aristotelian view on two fronts. First, he argues, if mixed elements do not preserve their form, then mixture is not a genuine union of elements – there would be nothing to be unified. The claim is not merely terminological: Sennert's point is that the Aristotelian view in effect turns mixtion into generation.[22] Second, Sennert takes experience to show that mixtures can be resolved into their component elements – which, he thinks, implies that the elements must preserve their forms and actually exist even while in the mixture.[23]

The claim that the elements and other *minima* retain their form in mixtion is, of course, crucial to Sennert's atomism. Because elements are individuated by their forms, a mixture in which the miscibles lose their forms is one in which they lose their identity as individuals. By preserving the forms of miscibles in mixture, Sennert has allowed the atoms to be components that exist in actuality in the mixture. At the same time, by insisting that a new supervening form comes into existence in mixtion, Sennert has provided an explanation of both the principle of unity of mixtures and the source of their new qualities.

A precisely parallel issue arises for Gassendi in the case of compound bodies in general, and although Gassendi's solution is more general, it is rather less successful. Gassendi's initial strategy is to do away with forms and make texture, internal corpuscularian structure, do all the work that forms did. Unfortunately, we shall see that texture is called on to carry more weight than it can bear. And in the case of living things, Gassendi recognizes the limitations of his initial strategy and reintroduces emergent powers and qualities to do significant explanatory work.

Let us now turn to Sennert's second question, whether only elements enter into mixtion. This amounts to the question whether all resolution to *minima* proceeds to the four elements. In first formulating his atomism,

[22] This is not an entirely fair criticism because one might hold that the elements preserve their forms potentially but not actually. I bracket out the issue of potential preservation because Sennert is like Gassendi in being fundamentally hostile to ungrounded potentialities.

[23] Sennert recognizes that it is disputed whether there are mixtures that can really be resolved into elements but thinks failure to recognize the truth stems simply from being a bad practical chemist. For instance, he cites the Coimbrans as denying that the vapor exuded when a log is burned is water and says that this claim is not only repugnant to Aristotle but also repugnant to the senses. Even though the liquor is not *pure* water, we can get pure water from it by distilling it. Sennert, *Hypomnemata Physica*, 127.

Sennert enlists Aristotle's remarks on the *minima corpora* at *De generatione et corruptione* 8 and the view of Democritus, Empedocles, and Epicurus that "atoms constitute the principles of natural things, from . . . which bodies are composed" as progenitors of his view.[24] Sennert's atoms are matter-form composites, and they come in a number of varieties. The atoms of the "*prima elementa*," earth, air, fire, and water, can mix and produce the "second genus of atoms" or "*prima mista*," salt, sulfur, and mercury.[25] These *prima mista* have *prima elementa* as their matter, and a new supervening form – that is, they are perfect mixtures in the sense described earlier. Similarly, perfect mixtures can be made taking the *prima mista* as the matter and ending up with a third-level atom possessing its own supervening form, such as an atom of gold. Like the *prima mista*, these third-level atoms have a certain stability that, along with the new form, makes them crucial for explanation. Just as certain "chemical operations" can break a gold atom down into *prima mista*, other chemical operations begin from an imperfect mixture of gold and silver and produce gold and silver atoms.[26] Sennert's higher-level atoms have a great deal in common with Gassendi's corpuscles, which are also imperceptibly small composite bodies with a certain stability and coherence that makes them useful for explanatory purposes. However, Sennert is far more explicit than Gassendi about the antireductionism of his account.

In contrast, Sebastian Basso, author of the widely read *Philosophia naturalis adversus Aristotelem (Natural philosophy directed against Aristotle)* denies that any new form can or must be taken on in mixtion and argues that all the qualities of mixtures can be explained in terms of the qualities of the components.[27] Basso begins in a typical manner by describing the views of various ancients, particularly Aristotle and the atomists, but he soon adopts the more radical tactic of criticizing the contradictions he sees as implicit within Aristotelianism.[28] Even though these criticisms are interesting, I shall focus on Basso's positive views on mixtion and, in particular, whether his atomism allows forms and qualities to emerge when atoms are combined.

[24] Ibid., 89 (for Aristotle) and 91 (for Democritus, Empedocles, and Epicurus).

[25] Ibid., 95 (for *prima elementa*) and 107 (for *prima mista*).

[26] As is standard, Sennert gives biological examples as well as chemical examples: For instance, the four elements mix to produce apparently homogenous substances like blood and bone, which then mix to produce organs, and so on.

[27] For an account of Basso's philosophy, its reception, and what is known of his life, see Lüthy, "Thoughts and Circumstances of Sebastien Basso."

[28] Basso, *Philosophia naturalis*, I.i.i.

Basso argues that mixture "is nothing other than the composition of elements that remain":[29] That is, *contra* Aristotle, the elements actually exist in the mixture and, *contra* almost the entirety of the tradition, mixtion is nothing beyond mere aggregation. He also makes the strong claim that the parts of the mixture are substantially different – a claim that Sennert could make only in the attenuated sense that formally distinct elemental atoms make up the matter of the formally homogeneous higher level atoms that are the parts of the mixture.[30] For Basso, who rejects forms, the qualities and powers of a mixture have to be explained in terms of the qualities and powers of its components alone.

Sennert and Basso, then, offer two importantly different types of atomism, and the difference has implications that reach far beyond the sphere of chemistry. Both types sharply restrict the qualities of the fundamental-level atoms. Only one type, however, allows that new forms and new real qualities emerge when fundamental atoms are combined. The other denies that possibility and insists that apparently new qualities of mixtures (and by extension all composite bodies) are reducible to the qualities of the fundamental atoms.[31] This taxonomy is helpful in seeing the range of options for Gassendi's atomism. The more programmatic opening sections of the *Syntagma* suggest the reductionist view. However, *arguably* by the time he discusses inanimate composite bodies, and *certainly* by the time he arrives at living animal bodies with generative powers, Gassendi abandons any such reductionist suggestions.[32] The *Syntagma* deals with the three realms of atoms, inanimate compound bodies, and living bodies in relative isolation from each other, and one of the tasks of this and the following chapters is to show how they relate.

Gassendian Atoms

Gassendian atoms are perfectly full, solid, hard particles, naturally and physically indivisible (1.258a). Their first two predicables, size and shape,

[29] Ibid., I.iv (section title).

[30] Ibid., I.i.vi.

[31] One could add a third view, the Lockean view that new forms and qualities are superadded by God to the appropriate composites rather than emerging naturally, but I am not aware of any self-styled atomists before Gassendi who held that view.

[32] I suspect that the same holds for Gassendi's account of paradigmatic chemical mixtures, although I have not looked at this in any detail. Given that Gassendi treats such living powers as nutrition and respiration in the chemical language of fermentation and the like, it would be odd for there to be any sharp separation.

are relatively simple. All atoms have one size or another, but there is some variation in size. Gassendi sometimes mentions explanations of macro-level phenomena that rely on the differing sizes of atoms – for instance, at points he suggests that the material soul's sensory ability can be at least partially attributed to its being composed of unusually small atoms – but this is relatively uncommon.

Gassendi is willing to allow the claim that all atoms have some size to imply that atoms are divisible in thought, and hence mathematically divisible. However, he insists on the irrelevance of this notion of divisibility to physics and does not make use of anything like the Epicurean doctrine of minimal parts. Although Gassendi is clearly aware of contemporary debates about indivisibility and infinity – he mentions, for instance, the frequently discussed problem of Torricelli's infinitely long solid[33] (2.264b) – he is unwilling to allow these mathematical problems to affect physics: "it is not permitted to transfer into Physics something abstractly demonstrated in Geometry" (1.265b).[34]

It is supposed to follow from the physical indivisibility of atoms that neither their size nor their shape can change. Atoms come in various shapes – hooked, sharp, rounded, pyramidal, and so on (1.268b). However, only rarely do appeals to the shape of atoms do explanatory work. Rather more often, Gassendi appeals to the shape of *corpuscles* or *molecules*, small composite bodies like the hooked corpuscles notoriously emitted from magnets.

From the fact that atoms contain no void spaces but are perfectly solid, it is supposed to follow that they are indivisible – and hence indestructible by natural means – and impenetrable (1.262b). A sponge can be penetrated by water because the water fills the holes in the sponge, but nothing can penetrate a poreless body like an atom. Impenetrability, then, differentiates body from vacuum; it also distinguishes the *concept* of body from the *concept* of vacuum. Gassendi tells us that the "anticipation or notion which we have of body is as having dimensions and being capable of resistance" or impenetrability, where having dimensions, of course, applies also to space (1.55a). Recall that the difference between space and body – that is, between incorporeal and corporeal dimension – was, in the first place, the distinction between dimensions that can be occupied and ones that are by nature full. Resistance and its approximate

[33] Mancosu and Vailati, "Torricelli's Infinitely Long Solid."

[34] In support of this, in the same passage Gassendi notes that Euclid himself granted that we should work with certain minimum angles in optics.

synonym, impenetrability, are thus criteria for being an atom and not merely powers of atoms.

The Weight of Atoms and Their Causal Powers

Atomic weight – *pondus, gravitas,* or *vis motrix* – is the most interesting of the features of atoms. Gassendi defines weight as follows:

> Heaviness [*gravitas*] ... or weight [*pondus*] ... is nothing other than the natural and internal faculty or force [*vis*] by which atoms move and go *per se,* or rather, [it is nothing other than] an innate, natural, native propensity to motion that cannot be lost, and a propulsion or impetus from within. (1.273a)

Later Gassendi adds that *vis motrix* is an "internal faculty" and an "inclination to motion" (1.273b). On his account, atomic *vis motrix* is the sole source of bodily activity within nature.[35] As such, it plays a central role in his system and is an important component of his attempt to revise Epicurean atomism so that it is acceptable to the faith. Let us step back, then, and see the role Gassendi takes bodily activity to play.

Like most early modern philosophers, Gassendi claims that ethics is the culmination of philosophy.[36] Now this is an entirely standard remark – one made even in textbooks like Eustachius's – but, in any case, ethics is certainly of central importance for him. Both ethical and religious concerns are manifest in Gassendi's arguments for the activity of matter, for he argues that only by making matter active can we preserve the secondary causation necessary for morality and religion.

For one thing, if there were no genuine causality save God's causality, then human beings could not be the cause of vicious and virtuous actions and thus could not be responsible for them (2.817a). Although the ultimate cause of morally relevant actions is the incorporeal human soul, those actions require corporeal activity as well. The soul's dispositions or intentions cannot, after all, issue in actions without the intervention of the human body.

Human freedom as a secondary cause is a special case, and Gassendi has more general reasons for thinking secondary causation is necessary as

[35] There is one somewhat problematic exception: the incorporeal human soul, which has causal efficacy over the human body in a way that, as we shall see in Chapter 10, Gassendi cannot fully explain.

[36] At some points, he also claims that happiness is the proper end of philosophy. Since Gassendi argues that virtue leads to happiness – whether in this life or the next – the claims are entirely consonant (1.3b).

well. He thinks that we experience the causality of created things in sense perception, endorsing Aquinas's claim that we know by sense that a body such as fire heats another.[37] Gassendi also suggests that we should read the words of Scripture "as they sound," and that a literal reading of the first chapter of Genesis – where "God commanded the Earth and Water to germinate and produce Plants and Animals" – shows that God has endowed creation with activity (1.493a; compare 1.487a).[38] Finally, he claims, again following *Summa Contra Gentiles*, 3.69, that it would detract from God's power and greatness if he did not confer some active power on created things (1.239a). None of these claims are elaborated or defended in any detail; Gassendi is simply mentioning, at various appropriate places, standard arguments for secondary causation.

Early on in the *Syntagma*'s *Physics*, Gassendi develops an account of the material and efficient principles of nature that he intends, among other things, to replace the common scholastic model of form, privation, and matter as the three principles of natural bodies.[39] The account revolves around two questions: *What is matter like?* And *what is the principle of activity within nature?* Or, more simply, *what kinds of things are secondary causes?*

This second question amounts for Gassendi to the question *what kinds of created things are efficient causes?* For Gassendi holds, at least in this portion of the *Syntagma*, that efficient causation is the only kind of secondary causation.[40] (We can, of course, usefully think of God's intentions for the created world as final causes, and such final causal accounts can often be provided more easily than efficient causal ones (3.358a ff).) Indeed, Gassendi treats the equation of efficient causation with causation in general as a "presupposition," stating: "It seems evident that the efficient cause, and the cause as such, are one and the same thing" (1.283a). The

37 Aquinas, *Summa Contra Gentiles*, 3.69.

38 As we saw in connection with Copernicanism, Gassendi is not entirely consistent on the subject of Biblical literalism.

39 See, for instance, Burgersdijck, *Idea philosophiae tum moralis*, 6.

40 As we shall see in Chapter 8, although this claim is put forward as a general truth, it may not be intended to apply to things produced by generation. Compare Osler, "How Mechanical Was the Mechanical Philosophy?" 191. Osler argues against the view that Gassendian matter is active, which is put forth by Bloch, *La philosophie de Gassendi*, 222 ff. Bloch, in turn, argues against Alexandre Koyré's charge that atomic activity cannot be reconciled with the "principle of inertia" (Koyré, *Études galiléennes*, 294-307). Bloch is correct to point out that there is nothing strictly inconsistent in holding that atoms move because of their activity while composite bodies have uniform and perpetual motion. However, the conservation principle he ascribes to reconcile the two lacks textual support and fits in badly with Gassendi's suggestions that atomic activity *underlies* the uniform and perpetual motion of bodies.

Exercitationes offered some argument for this thesis by arguing against the other three kinds of cause – final, formal and material – although, as we saw in Chapter 2, this argument often succeeded only by aiming at a highly simplified target. However, Gassendi's assumption that efficient causation is the only relevant sort of causation is legitimate in this context because he is concerned with causal relations between distinct bodies, which were traditionally thought to be efficient causal relations anyway.

Gassendi begins his account of the principle of efficient causality within nature by writing approvingly that Leucippus, Democritus, and Epicurus

> wished the Efficient Principle to be distinguished from the material principle only in virtue of a different way of considering them [*diverso respectu*], not in fact and by substance. For this is known from what was said earlier, namely, that the Atoms, which they said are the Material of things, are not considered to be inert and immobile, but rather most active and mobile. (1.334a)

He endorses a version of this claim that has been amended in three important ways, as follows.

(1) Epicurus held that all atoms move with a natural direction of motion downward – a view that notoriously requires postulating an uncaused atomic swerve to make collision possible. Gassendi objects both to the indeterminism of the swerve and to the assumption that space is directional in such a way that there is any one privileged direction of motion (2.837a). Thus, although Gassendi continues to use the traditional term *gravitas* for the weight or motive power of atoms, it is no longer apt: On his view, gravity is a product of corpuscularian emissions from the earth that hook onto and pull back certain composite bodies, so that motion in all directions is equally natural (3.487b ff).

(2) Epicurus held that all atoms move with the same speed, a doctrine that Gassendi holds cannot be justified given that we have no direct evidence as to atomic speed and that God could have created atoms with whatever speed he likes (1.335b). Rather, Gassendi writes:

> Nothing hinders us from supposing that some atoms are inert and that not all atoms are equally mobile . . . because all mobility in them was implanted by God as the author, some might have been created by God with outstanding mobility, some with moderate, some with little, some with none . . . [but on the other hand,] nothing hinders our supposing that . . . all atoms are implanted equally with the highest mobility. . . . One thing must equally be

supposed everywhere, namely, that however much mobility is innate in the atoms, that much constantly continues. (1.335b)

The last sentence of this passage bears some attention. It is open to either a strong reading, on which the mobility of each individual atom is conserved, or a weak one, on which the mobility of atoms in general is conserved but individual atoms can transfer motion. Thus, there is a question whether Gassendi held the Epicurean view that no individual atom can ever change speed, or merely a view on which the total quantity of atomic motion is conserved. I shall return to this issue at the end of the chapter, for to resolve it we need to better understand Gassendi's reasons for adopting a view of atoms as intrinsically efficacious.

(3) The final and perhaps most important amendment of the Epicurean account of atomic activity concerns the *source* of atomic activity. In place of the Epicurean claim that atoms are eternal and self-existent, Gassendi insists that atoms are created by God and are active because God created them as active, that is, instilled in them a *vis motrix* (or *pondus* or *gravitas*) at their creation: "It should be granted that atoms are mobile and active because of a force of moving and acting that God gave to them in his creation of them" (1.280a; compare 1.335b).

It is important both to my argument and to the physical consequences that Gassendi draws from his account of efficiency that Gassendi does not simply intend the claim that matter was put into motion by God: He intends the stronger claim that atoms contain within themselves a source of motion. However, the claim that the *vis motrix* of atoms is due to God's creation has been read as a denial of the genuine activity of matter. Margaret Osler argues that

> Gassendi believed that atoms are mobile and active because of the power of moving and acting that God instilled in them at their creation. If their mobility and activity were indeed innate, the dangers of materialism would be very real. Rather, he claimed, their mobility and activity function with divine assent, "for he compels all things just as he conserves all things."[41]

This is a difficult issue. On Osler's interpretation, Gassendi holds that "motion is imposed on atoms by God" (192) so that atoms are neither innately nor intrinsically moving. (Gassendi uses the term

[41] Osler, *Divine Will and the Mechanical Philosophy*, 191.

"innate" – *innata* or *ingenita* – but this of course is not conclusive.) Osler seems to think of Gassendi as one of a homogenous group of "mechanical philosophers" who worried that "active matter, insofar as it is self-moving, seemed capable of explaining the world without needing to appeal to God or the supernatural" – a danger that could be avoided "if matter were considered naturally inert and able to produce its effects only by mechanical impact" so that God was necessary as the source of motion.[42] Now, it is entirely correct that by insisting on God's role as the cause of atomic *vis motrix*, Gassendi is trying to mitigate worries about the atheism associated with Epicurean theories. But Gassendi – unlike later English corpuscularians – simply does not infer from this that we must disallow the activity of matter in order to avoid atheism. Rather, as we have seen, he argues that active matter is required in order to preserve secondary causation and thus to preserve religion. There is clear conceptual space for holding both that matter is genuinely active and that God must create and concur with material activity.

Osler's argument relies on texts that say that God is the ultimate source of atomic propensity to motion and must conserve the moving thing and "cooperate" with its "power of moving or acting" (1.280a). Taking these texts to imply a denial of the activity of matter runs together what should be treated as two distinct questions: first, the question of whether there is activity *in the created world*; and second, the question of whether this activity derives from and relies on God or is altogether independent. Consider the soul: It is not commonly taken as a barrier to the soul's activity that God created active souls and must concur with their activity. Nor was it generally taken to undermine hylemorphism that God is the ultimate source and preserver of the activity of forms. Claims that God conserved the created world and concurred with its activity were entirely standard in the late sixteenth and early seventeenth centuries.[43] If there are worries later in the century that allowing activity to creation will lead to atheism, they have not yet emerged in the 1640s; at least, neither Gassendi nor those writers he is arguing against demonstrate those worries. Let us, then, see what the accounts of conservation and concurrence lying behind Gassendi's account of the activity of matter are.

[42] Ibid., 178.

[43] See Freddoso, "God's Concurrence with Secondary Causes: Why Conservation Is Not Enough" and Freddoso, "God's General Concurrence with Secondary Causes: Pitfalls and Prospects." Freddoso identifies the occasionalist position that sixteenth-century scholastics like Suárez and Molina react against as the much earlier position of Gabriel Biel and Peter d'Ailly.

Conservation and Concurrence

The issue of occasionalism does not figure into Gassendi's account of activity explicitly. It is always a ground rule for him that whatever account of efficient causation one gives, it must preserve the genuineness of secondary causation. This ground rule was to be abandoned within a generation or two; hence, it may be helpful to see why Gassendi thinks that ascribing secondary causation to matter is theologically acceptable and, indeed, theologically necessary.

Scholars describe three more or less distinct pathways to occasionalism in the seventeenth century.[44] The first pathway starts by emphasizing the notion, common among seventeenth-century "new philosophers," that matter is inert, devoid of all special powers, and capable of moving from place to place but not of spontaneously producing movement. Thus, whenever a given body is moving, the initial cause of its motion cannot be that body itself (for it was assumed to be inert) and cannot be a different body (for if motion is understood as a mode of bodies, it is unintelligible to suppose it is transferred from one body to another). Hence by process of elimination the initial cause of every motion of a body must be God, and there is no genuine secondary causation: The collision of one piece of matter with another may be the occasion for God to put the second in motion, but it is not a real cause. The essential element of this first pathway is the inertness of matter, so Gassendi easily avoids this Cartesian or Malebranchian path to occasionalism.

The second and third suggested pathways to occasionalism depend on the doctrines of divine conservation and divine concurrence, respectively. Some philosophers have thought that continual creation leads directly to occasionalism in virtue of being incompatible with creaturely activity. It should be noted, of course, that continual creation goes back at least to Aquinas and was an extremely popular view, one that the majority of scholastics found entirely compatible with creaturely activity. Thus, one should not simply assume that Gassendi would have seriously considered the possibility that continuous creation implies occasionalism. Indeed, as far as I know, the debate he is intervening in prescinds entirely from

44 For instance, Nadler, "Doctrines of Explanation in Late Scholasticism and the Mechanical Philosophy," emphasizes both the first and the second. Clatterbaugh, *The Causation Debate in Modern Philosophy*, discusses all three. Freddoso, op./cit. identifies the path through divine concurrence as the route to the medieval occasionalism of Gabriel Biel and Peter d'Ailly and suggests that it is important for seventeenth-century occasionalism as well.

questions about conservation. This should be expected because the question at issue is the *locus* of secondary causation rather than its existence.

Conservation and its gloss as continual recreation are at issue in the debate between Descartes and Gassendi. Consider Descartes's Third Meditation claim that *it is evident by the natural light that conservation differs only by reason from creation.* Gassendi challenges this claim, asking, "how is this evident except perhaps in the case of light and similar effects?" (3.344b).[45] Descartes replies, with some irritation, that in denying "that we need the continual influx of the first cause in order to be conserved," Gassendi is denying "something that all metaphysicians affirm as evident."[46] Gassendi, who would surely have recognized the claim that we need the continual influx of the first cause as Thomistic, replies in equally irritated fashion that he is *not* denying that there is a continual influx of the first cause, God, into the created world (3.346a). Everyone agrees that the world "has nothing from itself by which it subsists *per se*" (1.323b). Rather, he is denying that this continual influx is equivalent to creation. Nor is he denying something that everyone accepts as evident, since the nature of conservation is disputed among the doctors of the schools (3.346a).[47]

However, this passage, like many from the *Disquisitio*, needs to be read with care. One cannot be sure that Gassendi is advancing a positive view of his own, rather than suggesting that Descartes has not ruled out a possible alternative. Indeed, at one point in the *Syntagma* Gassendi remarks in passing that "it is granted that conservation is nothing other than continual production" (1.485b). It is hard to know how to weight this claim against what is said in the dispute with Descartes. In general, texts from the *Syntagma* carry more weight than texts from the *Disquisitio.* On the other hand, the *Disquisitio* contains a series of articles on this issue, while the *Syntagma* claim is merely a note in passing – and one made in the context of an insistence on the genuine activity of matter in the form of

45 The relevant contrast is between effects like light and effects like heat. Sunlight persists only so long as the sun shines, but a boiled kettle remains hot even after taken off the fire.

46 Aquinas, *Summa theologiae,* 1 q.104a1.

47 Gassendi does not identify the dissent. However, Suárez, who himself argues that conservation "differs from creation only by reason of a certain connotation or implied negation – that is, it is only conceptually distinct from creation" (Suárez, *On Creation, Conservation, and Concurrence,* 120 (21.2.2)), cites the thirteenth- and fourteenth-century doctors Henry of Ghent, Gregory of Rimini, and Peter Aureole as refusing the merely conceptual distinction between conservation and creation. It is clear that Suárez takes this to be a minority view.

atomic motion at that. The *Syntagma* refers to conservation in a number of places, but it contains no real account of conservation. Gassendi's claims about conservation there are intended to help make his atomism acceptable to Christians. Their task is rebutting Epicurus, not constructing a workable theory of divine action.

At this point, we can draw two conclusions. First, Gassendi himself saw no reason to think that adopting a view of conservation as continual creation would lead to denying the activity of matter. What he objects to in Descartes's application of the doctrine is not any alleged implication of occasionalism but rather its use in a proof of the existence of God. Second, we have some reason to think that Gassendi is not particularly concerned with the proper theological understanding of conservation. Thus, given that he thinks there are acceptable theological authorities on which to base a refusal to identify conservation with creation, if he had been forced to deal with an alleged route from continual creation to occasionalism he might well have responded by denying continual creation rather than embracing occasionalism. However, this is pure speculation; I see no reason to think Gassendi entertained the possibility of such a link.

The third suggested route to occasionalism starts from the notion that God must concur or co-operate with the actions of creatures as well as conserving the existence of created things – a notion some writers before Gassendi, namely the medieval occasionalists Gabriel Biel and Peter d'Ailly, had found ultimately incompatible with genuine creaturely action.[48] Gassendi again assumes a rather underdescribed position. He discusses concurrence only, so far as I know, in the course of rebutting the Epicurean view that the gods are unconcerned with human affairs, but his characterization of general providence makes clear that both God and created causes have genuine causality in the case of any bodily action:

> Although authority and dominion are granted to God, the causes that he wishes to exist and to allow to act in their turn are not denied as a result of this. For it is his general providence that established the course of nature and permits it to be conserved continuously.... God is in fact supposed as the general cause of... all things. But moreover, particular causes are required... [that] are comprehended within a series of natural causes, which God permits to act in their turn. (1.326a)

As the light analogy suggests, Gassendi holds that God is the general cause of *actions* as well as creatures: "God cooperates with all secondary causes" (1.337b) and permits them to act. This is exactly the same terminology

48 Freddoso, "God's General Concurrence with Secondary Causes: Pitfalls and Prospects," passim.

used to describe how God cooperates with atoms' power of moving and acting (1.280a). By holding that both God and bodies are full causes of bodily effects, Gassendi falls squarely within the concurrentist tradition sometimes thought to derive from Aquinas and his *Summa contra gentiles* claim that

> It is not the case that the same effect is attributed to a natural cause and to the divine power so that it is effected partly by God and partly by the secondary cause. Rather, the whole is effected by both of them in different ways.[49]

Both God and creatures are genuine, immediate, and full causes of their effects. But what is it, on Gassendi's view, for there to be both a full, immediate general cause of a particular action and a full, immediate particular cause of the same action?

He gives us, as far as I can tell, nothing particularly helpful to go on here. Indeed, I can think of a number of reasons why Gassendi might have thought that it was not his job to explicate divine concurrence. First, he holds that we have little positive knowledge of the divine nature or its operations. We can only conceive of God, and hence of divine action, on a model with human action.[50] Such a conception of God is enough to ground knowledge and worship of God, but without the benefit of revelation it cannot enable us to explicate the divine nature.[51]

Second, to the extent he is aware of the debate between concurrentists and occasionalists, he takes the issue to be resolved in favor of concurrentism. Thus, Gassendi would not have taken himself to have any more need to explain the details of divine concurrence than other natural philosophers. All accepted that creation was active in one way or another. Concurrence is a topic that should fall within the boundaries of metaphysics, not physics, and Gassendi wrote no book of metaphysics.

Third, Gassendi has good rhetorical reasons to avoid giving any particular account of concurrence. He does not want his new, atomist philosophy too closely tied to controversial theological theses. Instead, he wants it to be acceptable to as many Christian readers as possible. His chief goal is to render atomism compatible with whatever we know must be true in religion and theology, and it would not help serve that goal to make any unnecessarily controversial claims in theology or metaphysics.

49 Aquinas, *Summa Contra Gentiles*, 3.70.

50 Here it is relevant that although Gassendi uses the traditional example of the sun conserving as well as creating light, it is not actually apt given his corpuscularian account of how light emanates from the sun.

51 See Chapter 10 for more on this point.

Fourth – and, I think, most important – Gassendi does not provide a detailed account of concurrence because worries about how conservation and concurrence can be compatible with creaturely activity are simply not part of the debate he is intervening in. Neither Descartes, nor the neo-Platonists Gassendi attempts to refute, nor the mainstream of late sixteenth- and early seventeenth-century Aristotelianism were concerned to challenge the genuineness of secondary causation.[52]

The Conservation of *vis motrix* and Atomic Motion

It is clear that Gassendi accepts *some* sort of conservation principle, either for atomic *vis motrix* or for motion itself. However, it is rather less clear what precisely is conserved, and a number of questions arise. Is the quantity of motion of each individual atom always conserved so that atoms can never change speed, but only direction, as a result of collision? Is it merely the quantity of motion in the universe as a whole? Alternatively, is it *vis motrix*, and not motion itself, that is conserved? If so, a parallel question arises: Is the *vis motrix* of atoms in general conserved, or individual *vis motrix*? Gassendi's language is less than clear. One definition of atomic weight suggests that the motion of atoms is conserved *because* their *vis motrix* is conserved:

> [Weight is] an innate vigor or internal energy . . . because of which [the atoms] are moved through the vacuum . . . in such a way that, because the vacuum is infinite and lacks any center, they will never cease from this motion of theirs, which is natural to them, but in every age will persist in this motion, unless other atoms or composite bodies lie in the way and they are deflected in another direction. (1.276b)

However, this passage does not make entirely clear whether individual atoms, or merely atoms as a whole, persist in their *vis motrix* and hence in their motion. Gassendi's claim that "there persists in things, in a constant manner, as much impetus as there was since the beginning" (1.343b) is similarly inconclusive. The absence of any account of transfer of impetus suggests that it is individual impetus or *vis motrix* that persists, but given

[52] Of course, worries about the compatibility of divine and creaturely activity become central topics within a few decades, as Cartesianism evolved. It is interesting to consider what happened in those decades to account for such a dramatic change – whether, for instance, the spur is something within Cartesianism itself (other than the common equation of conservation with continuous creation), some set of changes in the theological milieu of France, or simply the different intuitions of a new generation of philosophers.

that Gassendi does not pay attention to the collision of composite bodies either, the suggestion is pretty weak.

There are also more purely philosophical motivations for thinking that the persistence of the *vis motrix* of individual atoms implies the persistence of their quantity of motion. Our question is whether *vis motrix* is always realized in motion, or whether it can inhere in atoms as mere potential. Put in these terms, it seems that Gassendi would have to favor the first. He is generally suspicious of primitive potentials, and he provides no account of what would lead such a potential to become realized, if it is not always already realized. For instance, Gassendi's discussion of the creation of atoms has God creating them with *vis motrix* and, apparently as a result, already in motion: There is no evidence that God must do something other than create atoms with *vis motrix* to get them moving. Indeed, if having *vis motrix* did not entail being in motion, then Gassendi's whole account of the activity of matter would collapse. For we would need some *further* secondary cause of the actual activity of matter, one that is never in evidence in his discussions of secondary causation.

I suspect, then, that despite the obscurity of the texts, we are best off reading Gassendi as holding that individual atoms persist in their *vis motrix* and, as a result, in their motion. This is how Bernier understands Gassendi in his *Abregé* of the *Syntagma*.[53] No atom, then, ever changes speed, merely direction. Thus, even when a composite body is at rest, the atoms comprising it are continually moving in locked patterns of vibration. This last claim is genuinely Epicurean,[54] and Gassendi understands it as such, explaining that according to Epicurus "even in compound bodies atoms are continually moved" (1.279a).

However, scholars have, understandably, been somewhat reluctant to view atoms as always in motion, given that the texts are not conclusive.[55] Such a view has a number of unattractive consequences. It would eliminate transfer of motion; the only thing that impact could cause would be change of direction. (One might think such a consequence is in part

[53] Bernier, *Abrégé de la philosophie de Gassendi*, "Doutes," 2.300–1, mocks this view.

[54] Diogenes, *Lives of Eminent Philosophers*, 10.61. See also Lucretius, *De rerum natura*, 2.100–4 and 2.444–6.

[55] See, for instance, Brett, *The Philosophy of Gassendi*, 272 ff; Brundell, *Pierre Gassendi*, 84 ff; Bloch, *La philosophie de Gassendi*, 222 ff. Even though none of these authors assert conclusively that atoms do in fact move in the way concretions do, they all grant that the issue is troubling. A stronger claim – that Gassendi's physics is thus actually inconsistent – is made by Koyré, *Galileo Studies*, 278. Fisher, "Science and Scepticism in the 17th Century," 289, endorses this judgment, although it is not the main point of his analysis.

desirable, given that it eliminates ontological worries about transfer of motion, but Gassendi does not embrace it as such.) More worrying, it implies that atomic and corporeal motion obey entirely different rules. Gassendian composite bodies, as we shall see in the next chapter, obey roughly Galilean rules and change speed as the result of various factors. This seems entirely arbitrary, and it is hard to square with the way Gassendi treats micro-level phenomena as analogous to macro-level ones.

In the end, however, it is not so hard to see why Gassendi might be untroubled by the suggestion that there are radical discontinuitites between atomic motion and the motion of concretions. There are already a number of radical differences between atoms and composite bodies: Concretions are divisible but atoms are not; concretions have various occult and manifest qualities but atoms do not; the macroscopic world is continuous but the atomic world is not; and so on. Of course, in some of these cases the difference between the atomic and macroscopic levels is supposed to be explicable in terms of atoms, as in the case of divisibility. However, we shall see in the next chapter that Gassendi thinks that the "uniform and perpetual motion" of macroscopic bodies in a vacuum is explained by the underlying *vis motrix* of atoms. Indeed, at one point he claims, "all *vis motrix* that is in concretions comes from the atoms themselves" (1.343b; compare 1.384b). It will become clear that even "uniform and perpetual" motion requires some *vis motrix* or, as Gassendi tends to call it in the context of composite motion, impetus. This gives us at least the beginning of an explanation of the motion of concretions in terms of atomic motion.

We must also recognize that Gassendi was not particularly concerned with giving an account of the motion of atoms. (Had he been interested in atomic motion as an end in itself, it would be incredible that textual exegesis gave no conclusive answer to the questions just discussed.) Gassendi's accounts of macroscopic and atomic motion are generally provided in different sections of the *Syntagma* and in order to serve different goals. The account of atomic motion, I have argued, serves the goal of preserving secondary causation in the face of the incoherent or theologically unacceptable suggestions of his predecessors.[56] Gassendi's account of the motion of composite bodies serves a much narrower purpose, one

[56] Recall that Gassendi does not engage with Descartes's account of motion. Instead, he brushes it off as a nonstarter on the grounds that Descartes has eliminated any source of activity within bodies and thus will need to go back to an incorporeal principle like the world soul.

that is not connected to the foundational principles of his philosophy. It aims at explicating the physical causes of the phenomena of motion described by Galileo, and thus at vindicating Copernicanism as well as helping found a new physics of motion. In the next chapter, I look at Gassendi's account of the motion of composite bodies and the context in which it was put forward. Doing so will provide us with more insight into the relation between the atomic-level and composite-level descriptions of the world.

7

Bodies and Motion

Gassendian bodies are composites of atoms "woven together" with a certain texture. The notion of texture is central to Gassendi's atomism. In theory he relies heavily on the claim that the qualities and behavior of macroscopic bodies derive from the size, shape, and local motion of tiny particles, but in practice Gassendi almost always appeals to the notion of texture in providing explanations of particular bodily phenomena. This is one of the reasons his explanations are almost always qualitative rather than quantitative. Indeed, even in the one case where he provides a relatively detailed quantitative description – the proportion by which bodies in free fall accelerate – Gassendi makes clear that it needs to be grounded in an account of the textures of the relevant bodies. His concern with the new science of motion is above all to explain the underlying causes of motion, causes that must be described in corpuscularian, textural terms.

The program outlined in the initial books of the *Syntagma* suggests that texture is not something over and above the arrangement of atoms in space. However, it is unclear how we could derive anything that does all the explanatory work Gassendi needs textures to do from patterns of atoms moving in space. This chapter will examine both Gassendi's general account of texture and his account of the underlying causes of the motion of composite bodies. Treating these two issues together has a number of benefits. It will help us see how Gassendi's programmatic reliance on local motion turns into practical reliance on the rather loose notion of texture. It will make clear how Gassendi's underlying motivations differ from those of contemporaries like Galileo and Descartes. And finally, it will put us in a position to see how Gassendi would think the new explanatory resources he brings in to explain generation, life, and sensation – explanatory resources that go well beyond local motion or its products – are consistent with and indeed an integral part of his atomism.

I begin with the notion of texture in general, and then turn to Gassendi's account of gravity, which he takes to be a matter of corpuscular emissions. Because gravitational attraction is the result of relations between the textures of bodies and is also a cause of mathematically describable motion, gravity provides a bridge between explanations in terms of motion and explanations in terms of texture. With this bridge in place, I return to the relationship between atomic motion and the motion of bodies and, more generally, the relationship between Gassendi's programmatic atomist theses and his particular accounts of natural phenomena.

Composite Bodies

Ever since Boyle, Gassendi has been called a mechanist or mechanical philosopher. We tend to think of the category of mechanism as being defined both in terms of paradigm figures such as Descartes, Gassendi, Hobbes, and Boyle and in terms of paradigm doctrines such as the denial of forms, the explication of change in terms of local motion, and the inertness and homogeneity of matter. Now, it is contested what mechanism really amounts to and whether it is accurate or useful to speak of there being *one* thing, "the mechanical philosophy." If one understands mechanism as requiring being influenced by the discipline of mechanics or using it as a model, then Gassendi, who has no interest in mechanics and whose textural explanations are of a very different sort, will not plausibly count as a mechanist. He does use the metaphor of nature as a machine – the *machina mundi* – but no more so than the very different metaphor of nature as an army.[1] Thus, some find it preferable to call Gassendi's program an atomist program – a program of explicating the works of nature in terms of the qualities of atoms. Of course, in Boyle's taxonomy, atomism is one version of the corpuscularian hypothesis and thus a species of the mechanical philosophy.

Gassendi commits himself to explaining change in terms of local motion early on in the *Syntagma*'s *Physics*. For instance, he writes that

[1] In any case, the term "*machina mundi*" should not necessarily be thought of as having the mechanistic connotations it tends to suggest to twenty-first-century readers. For instance, in Ficino, Platonic Theology, the *machina mundi* is the body of the world soul. Fludd, *Utriusque cosmi maioris*, 17, similarly talks about the "*Macrocosmi machina*."

all qualities "are nothing more than certain local motions in which the *rerum principia* [atoms] . . . variously mix" (1.487b), and that "all changes terminate in qualities" (1.371b) so that all change can be understood in terms of the local motion of atoms. And he is clear that "there are in Atoms no qualities other than Magnitude, Figure, and Weight or motion" (1.336a). Thus, all change at the sensible level proceeds by transposing, adding, and subtracting atoms within particular composite bodies, processes that occur only as a result of the local motion of atoms (1.336a). Similar remarks are made in *De motu*, where Gassendi claims that qualities "appear to be nothing more than certain local motions, in which the *rerum principia*, although very subtle and insensible, variously mix" (3.487b-8a).

Gassendi takes his claim that the corporeal world consists solely of atoms with size, shape, and motion to raise the following question:

> [I]f it is true that the only material principles of things are Atoms, and that there are in Atoms no qualities other than Magnitude, Figure, and Weight or motion, as I explained above, how, I ask, may it happen that so many other qualities beyond these are created and exist in things? (1.366a–b)

The first stage of his answer is to invoke a further set of properties, the *eventa* or accidental properties of atoms. Intrinsic qualities cannot be changed while the atoms possessing them persist, but accidental qualities can change over time because they pertain not only to the atoms themselves but also to their spatial location. Thus, it is essential to any particular atom that it has the particular intrinsic qualities it does, and essential that it has some determinate accidental quality. However, no *particular* accidental quality is required.

Gassendi considers a few of the sets of atomic qualities that had been offered by ancient atomists. Aristotle reports that Leucippus and Democritus held that the essential qualities are shape, association or dissociation, arrangement (*ordo*), and position (*situs*) – a report that, for obvious reasons, must be mistaken. Eventually, he arrives at the view that he thinks was Epicurus', namely that position and arrangement are the accidental qualities of atoms. In a trope seen all the way from Aristotle to Boyle, Gassendi's initial discussion of position and arrangement uses letters as analogies for atoms: Position is the respect in which N and Z differ, and arrangement is the respect in which AN and NA differ. Gassendi uses the letter analogy to bring out the significance of positional difference:

> [J]ust as the same letters are different in respect of seeing and hearing when they are positioned differently . . . so the same atom positioned differently affects sense

in different ways, just as, if it were a pyramid, sometimes its edge might penetrate into the senses and sometimes it would apply itself through its base. (1.367a)

Next, consider arrangement:

[A]s the same two or many letters, when they proceed or follow in different ways, suggest different words to the eye and the ear and the mind . . . so the same atoms, transposed in different ways, can demonstrate very different qualities or *species* to the senses. (1.367a–b)

Gassendi does not give examples, although one might well take the example of the pyramidal atom as giving the appropriate flavor here as well.

Once he has described the accidental qualities of atoms, Gassendi goes on to give a taxonomy of physical changes. The first of three possible types of change involves the transposition of atoms within a body. Liquefaction and putrefaction are examples of this type. It is not entirely clear how we know that such changes involve only transposition; in his description of a rotting apple, Gassendi seems simply to *assume* that the apple does not gain or lose matter. The second and third types of change proceed by the addition (in the case of nutrition, growth, and moistening) or subtraction (in the case of distillation and precipitation) of atoms from composite bodies. It is clearer why Gassendi adopts this account. We see that animals eat various composite bodies and (other things being equal) grow larger over time, so it is natural to suppose that certain atoms have been assimilated. In any case, Gassendi notes that most observable changes involve two or even all three of these processes at the same time:

For example, when a thing liquefies, not only do its parts change position, but also some enter in from the fire, and some more subtle ones go out from the substance as it liquefies, and so on for the other changes. As a result, when one type of transformation is said to be appropriate for a single kind of change, it should be understood merely as the principal and predominant one. (1.371b)

Thus, the example of the rotting apple may be supposed to be simply illustrative, and not to imply a commitment about whether matter is lost or gained in putrefaction. Indeed, we shall see this kind of pattern all the way through Gassendi's explanations of particular corporeal qualities and behaviors. He gives an account of the texture a quality or behavior arises from, and at the same time makes it clear that this account is a suggestion rather than a substantive natural-philosophical doctrine.

Gassendi goes on to say that because "all changes end up as qualities" (1.371b), he will skip an explication of how particular changes work and simply explain what constitutes the various qualities. It is easy to pass by this remark, but it is extremely helpful for understanding the direction Gassendi's explanations tend to take. For the move from an account of change to an account of qualities substitutes a synchronic account for a more difficult to provide diachronic account: It allows Gassendi to avoid describing the interaction of composite bodies and just give static descriptions of their internal structure. I do not think any significant theoretical commitment lies behind this move. However, it is an important ingredient in Gassendi's transition from the programmatic claim that the corporeal world should be explained in terms of the local motion of atoms to his actual practice of appealing to textures.

Texture

But what exactly is a texture? Gassendi says rather less about this than the centrality of the notion demands. He never, for instance, gives a definition of texture. However, some insight into the notion of texture can be gleaned by looking at the various roles Gassendi needs it to fill:

1) Texture is a mere system of relations; the texture of a body is the arrangement of atoms composing it.
2) Texture must explain why bodies have the qualities and powers they do. A magnet, for instance, attracts iron in virtue of having a certain texture.
3) The texture of a body is its principle of individuation and principle of unity (at least for inanimate bodies).
4) As (3) suggests, texture is involved in explaining identity through change.

Thus, texture plays more or less the same role in Gassendi's atomism that form plays for Aristotelians. The crucial change is that, for Gassendi, all qualities of a body are products of its texture so that no qualities are real qualities in the technical sense.

5) Finally, since it will turn out that the motion of bodies follows very different rules from the motion of atoms, we need to appeal to texture to bridge the gap between atomic and macroscopic motion.

To understand the role this last and most problematic notion of texture plays, we need to look at one of its central occurrences in Gassendi's account of the motion of composite bodies. Gassendi uses texture to explain occult qualities, in particular the occult quality gravity, whose explanation in turn underlies his account of both accelerated motion and uniform and perpetual motion.

Much of Gassendi's account of the motion of composite bodies is found in the three letters *De motu,* which have four main concerns:[2] to argue against the Aristotelian distinction between natural and violent motion; to give an account of gravity as akin to magnetism so that free fall is explicable in terms of an external attraction; to explain the physical causes of the proportion by which bodies in free fall accelerate (in other words, the proportion of degrees of speed acquired to times elapsed); and to identify the physical causes of uniform and perpetual motion and Galilean relativity. Achieving this last goal relies on Gassendi's explanation of gravity as a product of corpuscular emissions rather than an intrinsic quality of bodies. It is important to Gassendi to deny that gravity is intrinsic to bodies because such a denial underlies his account of uniform and perpetual motion, which in turn underlies his defense of Copernicanism from the objection that the behavior of falling bodies on earth shows that the earth must be at rest.

Natural and Violent Motion

A distinction between two sorts of local motion, natural and violent, plays an important role in Aristotelian natural philosophy. The distinction can be sufficiently well understood for current purposes without delving into the ontology of motion in any detail, but a brief survey will be useful. In Aristotelianism, each of the four elements has a natural place: the two heavy elements, earth and water, at the center of the universe, and the two light elements, fire and air, at the sphere of the moon. Thus, things constituted of heavy elements or a preponderance thereof have a tendency to move downward toward their natural place. This tendency is not, of course, always operative – a stone sitting immobile on the surface of the earth is still heavy and thus still has a tendency to move downward – but when it is realized, natural motion downward occurs.

The cause of natural motion, importantly, is internal or intrinsic to the naturally moving thing. Gassendi's contemporaries commonly understood this as the basis of the Aristotelian distinction. Digby, for instance,

[2] Gassendi also develops an account of the parabolic path of projectiles. This has two main, related, claims. First, projectiles follow a parabolic path. Second, projectile motion has two components: the horizontal, uniform and perpetual motion given to it by the projector, and the vertical, decelerated or accelerated motion given by the projector or by gravity. I take it that most of what is new or interesting about Gassendi's account of projectile motion is in the account of the motion induced by gravity (that is, the component of free fall). Thus, I omit projectile motion entirely.

is concerned to demonstrate that "no body hath any *intrinsicall vertue* to move itself towards an determinate part of the universe"; later, he makes the same point in terms of "a naturall intrinsecall inclination" downward.[3] Rather, Digby argues, "the motion of every body followeth the percussion, of Extrinsecall Agents" so that it is "impossible that any body should have any motion naturall to it selfe" in the Aristotelian sense.[4] Nor was this understanding of the distinction peculiar to its opponents. Eustachius explains that a natural motion is one whose "principle of motion is innate [*insitum*] by nature in the mobile itself."[5] There were, of course, differences between scholastics in how exactly the internal principle of natural motion was to be understood, but we can prescind from such differences for present purposes.[6]

In contrast, motion brought about by an external mover is violent. Thus, projectile motion is a form of violent motion (although one that also involves the natural motion of the projectile toward the center of the earth). The case of projectile motion introduces a new problem. When the mover is in contact with the moved thing, there is thought to be little difficulty understanding how the mover's causality can operate. But what happens in cases where the mover ceases to be in contact with the moved thing after the initial moment – after the stone, for instance, leaves your hand? What, then, is the cause of motion? We shall return to this issue later, in the context of Gassendi's account of "uniform and perpetual motion." Here I simply want to point out that the causality of violent motion caused problems that the causality of natural motion did not.

Gassendi objects that the distinction between natural and violent motion has no particular basis in the world and no real use within physics. He starts by challenging the claim that natural motion has a cause internal to the moved body. Any initially plausible way to distinguish intrinsically from extrinsically caused motion, he argues, either counts all motion as natural or (almost) all motion as violent. For instance, one might construe natural motion as motion in accordance with the natures of the moved things. Then, because all moving things are constituted of atoms with an innate, God-given tendency to move, we will conclude that all motion is natural: "[C]ertainly, there is no motion that cannot be judged to be natural, insofar as there is none that does not derive from the

[3] Digby, *Two treatises*, 70 (Section 9.8) and 76 (Section 10.1).
[4] Ibid., 76 (Section 10.1).
[5] Eustachius, *Summa philosophiae quadripartita*, 112 (Section 1.4.7).
[6] For an account of some of these differences, see Wallace, *Prelude to Galileo*, 291 ff.

principles of things [that is, atoms], the nature of which their author willed to have incessantly an impetus by which they could move" (3.487b). A second, again plausible, way of drawing the distinction cuts against this and issues in the verdict that (almost) no motion is natural. For we could also construe as natural only the original motion and direction of atoms, with the result that virtually all motion will be violent: "[I]t appears that there is no motion that cannot be judged to be violent, aside from the primeval motion, insofar as there is no motion that happens except by the impulsion of one thing onto another" (3.488a). Atomic collisions occur all the time, and almost all atoms, particularly those trapped within the locked patterns of vibration that comprise composite bodies, are moving in the direction they are as a result of collision.

Gassendi concludes that the world could equally plausibly ground a number of inconsistent distinctions between natural and violent motion and thus that the distinction is not useful. This conclusion is also supported by his account of the acceleration of bodies in free fall, which explains the underlying cause of acceleration as something extrinsic to the moved body, namely gravitational attraction, rather than the body's own heaviness. Indeed, Gassendi's cosmology – on which neither space nor the universe has a center in any relevant sense – would also make the distinction problematic. Hence Gassendi abandons the distinction between natural and violent motion altogether and introduces a corpuscularian account of gravity to do the work that the innate tendency of bodies to move toward the center of the earth had done for previous philosophers.

Gravity

As has long been recognized, Gassendi's account of gravity played a key role in his transformation of the Galilean science of motion into a far-reaching natural-philosophical world view.[7] Gassendi's account of gravitational attraction as the cause of the accelerated motion of a body in free fall was, for instance, taken up by the young Christian Huygens in his 1646 *De motu naturaliter accelerato*.[8] For it is Gassendi's theory of gravity that implies the chief points on which Huygens's *De motu* differed from the Galilean theory of motion that inspired it.

[7] Koyré, *Galileo Studies*, 244–50.

[8] Galluzzi, "Gassendi and l'Affaire Galilée," 531–2, citing Christian Huygens, *Oeuvres complètes par la Societé Hollandaise des sciences*, 11.69.

Galileo tends to assume that heaviness is an intrinsic property of bodies. A well-known passage from the beginning of the *Dialogue Concerning the Two Chief World Systems* speaks of a body being "naturally capable of motion" in virtue of having "a natural tendency toward some particular place." From this, "it naturally follows that in its motion it will be continually accelerating."[9] However, by the end of the *Dialogue*, it is clear that talk of natural tendencies is just a way of describing our own ignorance:

> we do not really understand what principle or what power it is that moves stones downward, any more than we understand what moves them upward after they leave the thrower's hand.... We have merely... assigned to the first the more specific and definite name "gravity," whereas to the second we assign the more general term "impressed force."[10]

Thus, the fact that accelerated motion is natural is not supposed to imply the rule for the proportion by which bodies in free fall accelerate.[11]

Gassendi found it unsatisfactory for Galileo's otherwise impressive account of motion to rely on unknown physical causes. He was not alone in the sentiment that Galileo ought to have investigated the primary causes of the motion of bodies. Descartes, writing to Mersenne about the *Two New Sciences* in October 1638, complained that Galileo "does not stop to explain things fully. This shows that he has not examined things in an orderly manner, and that, without considering the primary causes in nature, he has simply looked for reasons for some particular effects."[12] I take "the primary causes in nature" to refer both to the divine activities of conservation and concurrence that ground a notion of force and to physical causes such as the vortices responsible for the fall of heavy bodies.[13] Although Descartes's conception of the relation between force and God is idiosyncratic, the claim that acceptable accounts of motion had to exhibit the underlying physical causes was made by parties from a spectrum of divergent theoretical orientations and by people who accepted the Galilean proportion as well as those who rejected it. The Jesuit Honoré Fabri, for instance, insisted that motion had to be explained in terms of causes consonant with metaphysical principles and offered a different proportion, which gave similar results in many common situations but did

9 Galileo, *Dialogue Concerning the Two Chief World Systems*, 20.
10 Ibid., 234.
11 Galileo's rule is proposition 2, theorem 2, corollary 1 of the Third Day of the *Two New Sciences*.
12 Descartes, *Oeuvres de Descartes*, 2.380.
13 For the first, see Garber, *Descartes' Metaphysical Physics*, 389; Des Chene, *Physiologia*, 426.

not rely on what he saw as the metaphysically unacceptable supposition that acceleration is continuous. For if an empirically adequate account of motion were to rely on metaphysically unacceptable assumptions, then it must be rejected for that reason. Hence, Fabri, who argued that time is composed of finite, discontinuous instants and hence that acceleration must be discontinuous, rejected the Galilean proportion.[14] Kenelm Digby, who notes that "we owe the greatest part of what is knowne concerning motion" and its conditions to Galileo, thinks that after establishing the conditions of motion "we shall do well . . . to enquire after the cause of it."[15] Digby thus concurs with Gassendi's claim that the causes of motion require investigating. However, he writes that "we do not conceive that [Gassendi] hath in his discourse [*De motu*] . . . arrived to the true reason of the effect we search into" (73) and offers a different explanation.

Not all the Parisian philosophers agreed that a science of motion ignoring physical causes was either incomplete or unacceptable. Mathematicians like Fermat and Roberval were not tempted by such a view, and Hobbes devotes much more attention to the geometry of accelerated motion in free fall than to its physical causes.[16] However, the great French debate concerning the nature and causes of acceleration in free fall that also involved Fermat, Huygens, Torricelli, and the Jesuits Fabri and Cazré centrally involved the causes of acceleration.

The well-documented example of Mersenne helps bring out the importance of physical causes to the question of free fall.[17] Mersenne's 1633 *Traité des mouvemens* considers three possible causes of the fall of bodies: positive and real heaviness; air pressure; and terrestrial attraction. He rejects the first two, and says that terrestrial attraction would work, but "one can invent various other reasons" that would work equally well. Thus, "it is sufficient to explain the phenomena of nature, since the human spirit is not capable of possessing its causes and principles."[18] However, by 1647's *Novarum observationum tomus III,* Mersenne had moved in a more skeptical direction and no longer maintained that the Galilean proportion of acceleration was precisely right. There is some debate about why Mersenne moved away from the Galilean proportion, but it is agreed that he became concerned that no explanation of the physical causes

[14] Galluzzi, "Gassendi and l'Affaire Galilée," 523–4.
[15] Digby, *Two Treatises,* 63.
[16] Hobbes, *Elementorum Philosophiae Sectio Prima De Corpore,* 127–42 (3.16).
[17] Palmerino, "Infinite Degrees of Speed," passim; Dear, *Mersenne,* 208–22; Galluzi, "Gassendi and l'Affaire Galilée," 510.
[18] Palmerino, "Infinite Degrees of Speed," 270–1, citing *Traité* 24.

of gravity could account for the fundamental Galilean principle that a body in free fall passes through "infinite degrees of slowness."[19] He thus concluded that the physical causes of gravity are entirely unknown, and that the strongest claim we can legitimately make is that Galileo's proportion applies to heavy bodies falling the relatively short distances we observe.

Although the extent to which Descartes's account of gravity contributed to Mersenne's change of heart is disputed, it is certainly important. Descartes wrote to Mersenne that "weight is nothing other than terrestrial bodies being actually pushed toward the center of the earth by the subtle matter."[20] This is the central element of his vortex theory of gravity.[21] He took this theory to imply, first, that a body with less subtle matter in its pores will fall faster than a body of equal size containing more subtle matter; and second, that (since the initial push of subtle matter must put the body in motion with some finite speed) acceleration cannot progress through infinite degrees of slowness.[22] The second point, which seems to hold true for *any* contact-action theory of gravity, is what Mersenne was most likely worried about. I return to the issue of the continuity or discontinuity of motion accelerated by gravity later because, although it is not an issue in Gassendi's original *De motu* account, it becomes important in the later *De proportione*.

Gassendi thinks the physical causes of gravity *can* be identified, and even offers an account of them. This shows how far he has moved by the 1640s from the "mitigated skepticism" he originally shared with Mersenne. Gassendi begins from William Gilbert's claim that "the globe of the earth is nothing other than a great magnet" (3.487b).[23] Although Gassendi does not go so far as to endorse the claim that gravity and magnetism are the same, Gilbert's account of the magnetism of the earth is important to him because it suggests a physical mechanism for gravity on which it is extrinsic to bodies. Moreover, Gilbert's account has the great advantage, from Gassendi's perspective, of being explicitly Epicurean in

[19] Galileo, *Dialogue Concerning the Two Chief World Systems*, 28. For the debate, see Dear, *Mersenne*, 215 ff, and Palmerino, "Infinite Degrees of Speed," 513.

[20] Descartes, *Oeuvres de Descartes*, 3.9–10 (January 20, 1640).

[21] Ibid., 8.213 ff (*Principles* 4.23–7).

[22] Ibid., 3.10 ff (to Mersenne, January 20, 1640).

[23] I do not intend to claim that Gilbert is the only, or even the main, influence on Gassendi's conception of gravitation. Rather, I emphasize the line of thought that I take to have been inspired by Gilbert because it fits in well with my overall account of textural explanations of occult qualities in Gassendi.

inspiration. The obligatory history of magnetology Gilbert provides as an introduction to his own view claims that

> Epicurus thought that iron is attracted by a magnet just as straws by amber: and he added that the reason is that the atoms and indivisible bodies that flow from the stone and the amber are fitted in shape to each other so that they easily cling to each other.[24]

Thus, Gilbert says, when the magnetic particles rebound, they bring the iron back with them. Static electricity works similarly. For Gilbert, gravitational or magnetic attraction and electrical attraction are distinct phenomena with different causes, only the first of which relies on material effluvia; however, Gassendi uses his principle that all corporeal action requires contact to argue that magnetism, gravity, and electricity must *all* be conveyed through material effluvia.[25]

For Gilbert, the earth (and indeed all celestial bodies) is magnetic.[26] He takes this to imply an explanation of the tides in terms of the moon's magnetic attraction.[27] More to the point, he also takes it to imply that a stone dropped from a tree on a moving earth will fall directly below the point from which it was dropped, despite the intervening motion of the earth. The stone remains within the "sphere of influence" of the earth – which, Gilbert claims, refutes Tycho's objection to the earth's motion.[28] This is a note in passing; Gilbert is not so much concerned to explain why bodies fall to earth as to show that they would fall in the same way whether or not the earth moved. However, Gassendi takes up Gilbert's account to explain why heavy bodies fall.

24 Gilbert, *De Magnete*, 61 (2.3). Compare Lucretius, *De rerum natura*, 6.1002 ff, who refers to both the particles emanating from the magnet and the motion of air displaced by particles rushing out from the iron.

25 The reason Gilbert thinks electrical attraction must operate through a "primary vigor" or "primary and radical and astral form" rather than material effluvia is that electrical attraction is nullified if a piece of marble is placed between the two bodies, although gravitational and magnetic attraction are not. Gassendi suggests that the difference may be due instead to the differing shapes of magnetic, electrical, and gravitational corpuscles. Perhaps electricity corpuscles have the wrong size or shape to pass through the void spaces or pores in marble.

26 In fact, on Gilbert's account, matter in general is magnetic, although its "magnetic form" is often damaged so that its power is nonapparent. *De Magnete* 117 (Section 2.4). Gassendi concurs with the claim that all heavenly bodies most likely exhibit attractive power, although he does not extend this to matter in general. Neither writer develops the suggestion.

27 Gilbert, *De Magnete* 86 (Section 2.1).

28 Ibid., 229 (Section 6.5).

Gassendi justifies his claim that the heaviness of bodies is not innate but derives from the earth's attraction by appeal to examples of ordinary magnets:

[P]ick up and hold in your hand a small plate of iron of a few ounces. If then some very strong magnet is placed under your hand, you will feel a weight not of ounces but of some pounds. And because you admit that this weight is not so much innate in the iron as impressed from the attraction of the magnet placed under the hand; thus, where it concerns the *pondus* or *gravitas* of a stone or other terrestrial body, it can be understood that this *gravitas* does not so much belong to a body of this sort from itself, but from the attraction of the earth placed under it. (3.508a)

Such attraction, like magnetic attraction, operates through direct corpuscularian contact. Because Gassendi insists that "to attract is nothing other than to impel toward oneself by means of an inward-curving instrument" (3.497a), he thinks of gravitational attraction as equivalent to action by impact. Modeling attraction on impact will raise problems later because it suggests – *contra* Galileo – that acceleration is a discrete process. I return to this problem later.

In the case of a stone falling toward the earth, each corpuscle in the stone is linked to the earth by thin chains, also composed of corpuscles (and in turn of atoms). These chains or "magnetic rays" pull the stone toward the earth. One advantage of this account, Gassendi emphasizes, is that it explains the recent discovery that all bodies fall the same distance in the same period. If we consider two falling bodies, one huge and the other tiny,

the huge one is attracted by many little strings, but has many more particles that need to be attracted. Therefore, it happens that the *vis* [force or power] and the *moles* [mass or size] are proportional. And the proportionality [of *vis* and *moles*] in both cases is such that it suffices for carrying out the motion in the same time . . . one will reach the Earth hardly later than the other . . . since in the same way there is a proportionality so long as just as many little strings are attached to just as many particles. (3.495a–b)

(One will reach the earth "hardly later" than the other, rather than both arriving at the same time, because the effect of air resistance depends on size and relative weight.) Gassendi also embraces a second consequence, one whose desirability is not so obvious:

just as the power of a magnet is diffused in a sphere, with the result that corpuscles springing out in rays become rarer as they proceed further out, and so, being less dense, attract less, and as they pass through a certain distance become unequal to attraction and at last become nothing, in the same way the globe of the earth diffuses its *vis attractrix* in a sphere, with the result that the rays of corpuscles emitted at last become utterly rarified and from a certain distance cannot attract a stone. (3.494b)

The point that terrestrial attraction theories of gravity imply that gravity diminishes with distance from the earth had been made by Mersenne in the *Traité*. Gassendi accepts the point but does not see this as a problem. Given the great size of the earth and the relatively small distance we observe bodies falling through, we can disregard the diminution of gravitational attraction for practical purposes. However, the diminution of gravity with distance allows Gassendi to envision situations in accord with the order of nature in which bodies are not affected by gravity at all. The possibility of such situations is, as we shall see, central to his account of uniform and perpetual motion. Moreover, although Gassendi does not emphasize this point, given his account of gravity, he must think that the Galilean proportion of acceleration is merely a useful approximation. This helps us see why Gassendi continues to emphasize the Epicurean point that mathematics does not precisely describe the world even after offering his superficially Galilean account of motion.

A third consequence is, perhaps, worse. The account explains why some bodies are heavier than others: Bodies composed of more corpuscles are pulled by more chains.[29] Now, this explanation only works if it is assumed that the weight of bodies is proportional to the number of corpuscles they contain. Such an assumption is at best ad hoc. Because there are myriad types of corpuscles, each composed of different numbers of atoms with different sizes, shapes, and motive powers, it is entirely implausible that weight is straightforwardly proportional to the number of corpuscles. Although the account of gravity is the place where Gassendi most fully develops his attempt to combine matter and motion theory, the two do not really support each other as much as one might hope.

The account of gravity is both a crucial ingredient in Gassendi's account of the motion of macroscopic bodies and an important case of a textural explanation of occult qualities. I shall consider the textural explanation of occult qualities like gravity first because this will allow us to see how Gassendi's account of the causes of Galilean motion is related

[29] Westfall, *Force in Newton's Physics*, 109, makes this point in terms of atoms rather than corpuscles. This cannot be right because magnetic rays are one corpuscle thick, not one atom thick (3.495a) and hence must hook onto corpuscles rather than atoms.

to the basic principles of his natural philosophy. Seeing the natural philosophy in which Gassendi's account of the motion of composite bodies is grounded will allow us to make sense of various troublesome features of that account.

The Textural Explanation of Occult Qualities

Gassendi distinguishes qualities pertaining to individual atoms from those pertaining to concretions. The latter in turn are subdivided into two categories, "those pertaining to the senses" and "those commonly called occult" (1.375b). Occult qualities include sympathy and antipathy (such as the natural antipathy of sheep for wolves); magnetism and static electricity; the healing power of the weapon salve; and the poisonous glance of the basilisk. Gassendi offers explanations in terms of corpuscular emissions for all of these. He also notes that there are some alleged occult qualities – such as the influx supposed to run from celestial bodies to human beings that judiciary astrology relies on (1.452a) – whose effects are never evident and that we should thus dismiss altogether. Indeed, sometimes Gassendi suggests both ways of dealing with occult qualities at the same time, as when he writes that either the basilisk emits deadly rays of corpuscles from its mouth or its deadly power is just a fable (1.453b).

What makes qualities such as magnetism, electricity, and the basilisk's toxicity occult? Originally, occult qualities were those that are occluded or hidden from sense. Thus, they are observable by their effects alone, in contrast with qualities such as color, which were traditionally taken to be directly sensible. (For Gassendi, however, sensible qualities – at least if understood as textures rather than the appearances they produce – are also only observable by their effects.) The Aristotelian terminology of occult qualities also suggests that the occult qualities are in some sense mysterious or inexplicable. This is not necessarily intended pejoratively; Aquinas said that the attractive power of the magnet is beyond human understanding, and Gassendi at one point defines occult qualities as those "which are considered to have unknown causes" (1.449b).

The term "occult quality" is used in chemistry as well. Again the manifest and the sensible are equated, and the occult is thought of as hidden, either in virtue of incorporeality or in virtue of being located inside a corpuscle. This sense of hiddenness as corporeal interiority, in opposition to what is manifest on the surface of a body, is also at work in Gassendi's talk of the "hidden natures or essences" of things. But Gassendi does not have the same patience with talk of the unknowability of qualities. For qualities should not be unknowable in themselves, and if they have unknown causes, it is only because essences are unknowable.

Gassendi's strategy for explaining the underlying causes of occult qualities and dispelling the air of mystery that surrounds them requires satisfying three principles: first, the "perspicuous principle" that "nothing acts on a distant thing" (1.456b); second, that "there is no effect without a cause," (that is, no *event* without a cause); and third, that "nothing acts without motion" (1.450a). These principles, taken together, have a ring of mechanist orthodoxy to contemporary readers, who easily assume that Gassendi's principles are prefatory to an account of occult qualities in terms of local motion. But a reader who expects this would be disappointed, as the case of static electricity will show. Consider a passage where Gassendi is explaining why amber attracts small pieces of straw when you rub it:

> [I]nnumerable little rays are emitted like tongues from this sort of electric bodies. These rays insert their ends crosswise into the pores of light bodies, filling them, and snatch them up. As the rays are pulled back, they carry the light bodies back and hold them. (1.450b)

Gassendi adds that the little rays "leap forth because of the corpuscle's heat" and that they are brought back not spontaneously but again by rubbing (1.450b).

Margaret Osler discusses Gassendi's account of the amber effect, pointing out that he leaves unanswered why the same rubbing should make the cords first fly out and then fly in.[30] She also points out a structurally similar problem concerning gravity, which was first noted by Richard Westfall: What makes the chain of particles connecting the falling body to the earth reel itself back into the earth? Worse, what makes the chain reel itself back in *with acceleration*? Osler thus suggests that Gassendi has simply moved the problem of explaining electrical and gravitational attraction from the macro level to the micro level, with the result that his physics contains unintentionally nonmechanist elements.

Is this the right conclusion to draw from such cases? Gassendi's account of electricity respects the principle *nothing acts on a distant thing*, but it has rather more trouble with the other two principles. Because he has no real account of why the rays are pulled back in, he apparently posits

[30] Osler, "How Mechanical Was the Mechanical Philosophy?" 433–4.

an event without a cause. And because rubbing the amber is supposed to cause the rays to shoot out in the first place, there does not seem to be any motion left to explain why the tongues are pulled back in. However, the existence of such *lacunae* only shows that Gassendi is relying on fundamental qualities of matter beyond size, shape, and motion *if* we assume that his accounts of electricity and gravity are intended to be complete. We must, then, ask whether Gassendi is denying that any further explanation of how gravitational rays retract is needed, or whether he simply cannot yet provide such an explanation.

Gassendi does not make it explicit here that he is merely providing the best preliminary sketch he can. However, the suggestion can be found in many other places in the *Syntagma*. For instance, he insists that general maxims like the three principles just discussed are easier to come by than particular explanations that satisfy them:

> concerning general causes there is no one who does not proceed *more Geometrico* – "There exists nothing which is not made from matter; nothing without motion; and nothing which is not made by an agent." But when it must be said more specifically what matter, what motion, what agents and the like, there is no one who does not stumble and stammer like a little child. (3.362b; compare 3.493a)

Indeed, Gassendi's accounts of space, time, and atomic motion are put forth with much greater certainty than any explanations of particular phenomena. Thus, his textural explanations should generally be seen as intended merely to make a corpuscularian explanation *imaginable*, without carrying any great commitment to truth or completeness. Providing textural explanations has programmatic and rhetorical significance even where it is not clear how seriously the explanation is intended. The case of the basilisk – who either emits deadly corpuscles from his mouth or is a mere fairy tale – is a nice example.

Motion in Free Fall

In what has come to be known as the "second Galileo affair," much debate arose in the wake of the French reception of Galileo's account of free fall. The problem of free fall was not, of course, first posed by Galileo. Indeed, Gassendi had been interested in free fall – and in particular the proportion by which acceleration occurs – when he talked with Beeckman in July of 1629.[31] However, it is not clear how deep this interest ran. Although Gassendi was certainly impressed with Beeckman, as far as we know he did no further research on free fall until the late 1630s, when he was spurred by reading Galileo's *Two New Sciences.*

Two issues were central in the debate triggered by Galileo: the proportion by which bodies in free fall accelerate[32] and the underlying physical causes of such acceleration. Although Galileo himself focused almost entirely on the first, the second became crucial in French discussions concerning Galilean accounts of motion in the 1630s and 1640s. One dominant view as to the proportion of acceleration was Galileo's view that a falling body that traveled through one unit of space (from now on, one space) in one unit of time (from now on, one time) would fall three spaces in the second time, five in the third, seven in the fourth, and so on (so that distance fallen is proportional to the square of time in fall). Two other proportions were advanced. Fabri argued that the sequence of new degrees of speed followed the natural number series rather than the odd number series (so that a body that fell one space in one time would fall two in the second, three in the third, and so on), and Cazré contended that degrees of speed added doubled from moment to moment (so that a body that fell one space in the first time would fall two in the second, four in the third, and so on).[33]

Gassendi endorsed Galileo's odd-number sequence throughout his writings on motion. He devotes no real attention to establishing that this proportion is correct and, indeed, writes as though we literally *observe* the proportion. (This assumption is somewhat surprising because, as Fabri and others pointed out, the unaided senses cannot judge distance and time sufficiently precisely.[34]) His concern was to provide a causal mechanism that would yield that proportion, and, because he changed his mind, he ultimately proposed two different ones. I shall consider both mechanisms and why Gassendi rejected the first. This will provide us with

[31] Beeckman's journal account of this meeting can be found in Beeckman, *Journal tenu par Isaac Beeckman de 1604–1634*, 123–4. Compare Peiresc, *Lettres de Peiresc*, 4.198–202.

[32] "The" proportion, because most of the debate assumed, with Galileo, that acceleration in free fall is uniform (that is, that the motion of a body in free fall "adds on to itself equal momenta of swiftness in equal times"). Galileo, *Two New Sciences*, 154. Descartes's vortical theory, on which the acceleration of a falling body slows as its speed approaches the speed of the subtle matter pushing it, is the most notable exception.

[33] Palmerino, "Infinite Degrees of Speed," 296–7, notes that although Galileo's proportion gives the same results no matter what units of time and space were used, the other two do not.

[34] Dear, *Discipline and Experience*, 141.

an illustration of Gassendi's much-maligned mathematical competence, at the same time as illustrating the crucial but confusing role of continuity in his physics.[35] So doing will also help us better understand the relation between atomic-level and bodily-level explanations in Gassendi's physics.

In the first letter of *De motu*, Gassendi holds that the odd-number sequence cannot be explained merely by the acceleration resulting from gravitational attraction but needs a further cause as well – namely, the impact of air on the falling body (3.483a–b). He provides an interesting argument for this. It relies on a thought experiment involving a stone in a void, which is initially at rest because there is no cause for it to move in any particular direction. Suppose "that the body of the air is brought back by God in such a way that it surrounds the stone so that no attractive force from the earth can come to it" (3.496b). The stone will still be at rest because there is no cause impelling it from one part of the motionless air to another. Motion is introduced at the next step:

> [C]onceive that not the air, but only the effluvium of magnetic rays from the earth, is brought back; and you will discern why it is that the stone can be driven away from rest and attracted towards the earth. And in this way, the *vis attractrix* of the earth can be the cause that first brings motion in the stone. Now conceive that the air is brought back at the same time along with the effluvium; and you will discern that the stone can still desist from rest and be attracted towards the earth,because even if the inferior air (that is, the air under the stone) resists somewhat, on account of a certain bodiliness, nevertheless the attractive force is so strong that it suffices for moving the stone and overcoming the resistance of the air. (3.496b)

How *fast* will this motion be? To answer this question, Gassendi first supposes a body moving through a vacuum. In such a situation, he argues, once the body is put into motion by impact, "it will clearly be moved equably and with a perpetual motion unless an obstacle occurs" (3.496b). (I return to this claim in the following section.) Now conceive that

> while it is moved in this way it is struck by a similar blow; then, because the preceding motion is not destroyed, two motions will be united into one, which is faster by two than the first, and indeed in the same way will be equable and perpetual. Conceive that [the body] is struck again by a third similar blow, then because the earlier motion, composed as it were from two degrees [of speed], perseveres, there will occur a coalition of motions or degrees, that is, there will be a motion that is three times faster than the first. (3.497a)

[35] See Koyré, *Galileo Studies*, 245, esp. n. 21. Koyré is referring to Gassendi's initial "misunderstanding" of the Galilean proportion of acceleration.

And the gravitational attraction of the earth has to work the same way, on Gassendi's account, for "to attract is nothing other than to impel toward oneself by a curved instrument" (3.497a). Acceleration is thus a matter of a constant series of blows, one at each new moment. Gassendi takes this series of blows to yield the sequence of acceleration 1, 2, 3, 4...

The key here is that Gassendi thinks the chain connecting the earth to the falling body continues to exert its attraction at each moment because it remains in physical contact with the earth. However, the natural number series this yields is not the one that Gassendi recognizes as correct and that he says has been "seen" to be true (3.497b). Thus, we need an explanation of the stone's additional acceleration. He suggests the following:

> [B]ecause the air that presses around [the stone] presses downward but can never enter into the space below the stone unless that space is left empty, the air, springing back from the sides, comes together into the place above the stone and invades the space that has been left empty. And since this invasion cannot happen unless the air rushing in urges the stone onwards, therefore in the second moment two new blows occur, one from the Earth, which continues to attract, the other from the air, which begins to urge [the stone] onwards. Moreover, three degrees of speed result from the joining of these two motions or degrees of speed, which, along with the first [motion], are not destroyed but persevere. (3.497b)

The added effect of the air is very similar to traditional antiperistasis explanations, save that Gassendi replaces the horror of the vacuum by air pressure (*pressio aeris*) just as he does in the barometer cases.[36]

A few years later, in *De proportione*, Gassendi acknowledges that this explanation of the causes of acceleration was wrong. His reasoning had gone astray because

> it imprudently allowed velocities to be considered as spaces. For I did not attend enough to the fact that the degree of velocity acquired in the first moment remains intact in the second, with the result that it is able to go through two spaces – I considered [that degree of speed] as if it were able to go through only one space. Thus when I saw that in the second moment [the stone] went through three spaces, I promptly thought that just as one degree [of velocity] remained to pass through one [space], so in the meantime two more needed to be acquired to pass through two more spaces. (3.621b)

Thus, "velocity continuously [*continuo*] increases from the beginning of motion" (3.355b–556b). The continuous increase of velocity is what

[36] Gassendi notes in *De proportione* that the greater density, and hence greater pressure, of the air nearer the earth should be relevant as well, although he does not quantify its effects.

explains how gravity alone can yield Galileo's proportion. Each degree of speed acquired in a preceding moment remains intact in the next, where "it acts twice as much because of its constancy" (3.608b).[37] Gassendi has now distinguished between two things – the final speed the stone has at the end of each moment and the average speed it had throughout the moment – that were confused in *De motu*. Thus, the air-pressure version of antiperistasis Gassendi had used in *De motu* is no longer necessary to get the correct, Galilean proportion.[38]

Gassendi's attitude toward his mistake is interesting. Although his account of the effect of air pressure was relatively detailed and definite, he does not appear to be embarrassed by having to retract it. Instead, by showing how he fell into error, he manages to suggest that even competent opponents may equally have fallen victim to error. Gassendi obviously thought that an important rhetorical point was made by recounting his discovery of the mistake in his original two-factor theory and his subsequent switch to a model on which gravitational attraction alone was the cause of motion: In the *Syntagma* he explains both theories, and his reasons for abandoning the first, in detail (348b–53a). But his change of heart is also interesting as evidence that explanations in terms of textures of individual bodies have an entirely different epistemic status than the basic atomist hypothesis. The claim that gravity, combined with air pressure, yields the Galilean proportion describes *one* possible underlying physical cause; the claim that gravity alone does so describes another. Gassendi came to think the second was better because there were intrinsic problems with the air pressure account, but this does not show he is committed to his second explanation being either complete or demonstrative.

De proportione puts much less emphasis on physical causes than *De motu* and, in that sense, is a diversion from attempts to bring theories of motion and theories of matter together. But it does not *abandon* the search for physical causes by any means. Gassendi initially thought of gravity as impressing discrete degrees of speed at discrete moments. This way of thinking of the effects of attraction is natural on a model on which

37 Palmerino, "Infinite Degrees of Speed," 309, points out the centrality of this claim to understanding Gassendi's shift.

38 Gassendi adds that air pressure would not really have provided any extra impulse in the direction the stone was traveling in because "air does not only resist from above . . . but also from below" (3.622b). Moreover, if air pressure were playing a central role, then it would be difficult to explain why bodies with different weights fell with the same proportion of acceleration (Ibid.).

attraction is simply inward impulsion, a series of little collisions. However, even though the assimilation of attraction to impact suggests a discrete process, in other ways Gassendi's corpuscular model suggests a continuous one. The chain of corpuscles constituting gravitational attraction is always in contact with the falling body, so – on the assumption that time is itself continuous[39] – gravity should operate in a continuous manner. Gassendi's shift from thinking of acceleration in free fall as discrete to thinking of it as continuous does not so much require finding a new physical cause as understanding the old one in a new way.

That Gassendi's explanation of the underlying physical cause of accelerated motion in free fall can be made compatible with either the natural-number *or* the odd-number rule shows the rather limited extent to which he grounds facts about the motion of composite bodies in his atomism. His account of gravity is one of the cases where Gassendi most explicitly tries to tie matter and motion theories together, but it is not an entirely successful one. We shall see that the same may be said for the case of uniform and perpetual motion.

Uniform and Perpetual Motion

Although there had been much dispute over what keeps a projectile in motion when it is no longer in contact with its projector, Buridan's impetus theory, suitably modified, had become the consensus position of early

[39] This assumption is somewhat troublesome. The account of *De proportione* seems to require that space and time be continuous, and the ontology developed in books 1 and 2 of the *Syntagma*'s *Physics* does not seem to allow for space and time to be discrete. Because the only real divisibility for Gassendi is physical divisibility, incorporeal entities like space and time cannot be divisible. And even though the view of space and time as composed of *minima* has Epicurean credentials, Gassendi has obviated the need for space *minima* by abandoning the doctrine of atomic minimal parts. However, Gassendi does sometimes allude to an atomism of space and time. 1.341b–2a is by far the clearest case: Gassendi talks about differences in speed being due to "the intermixture of fewer or more particles of rest" and, echoing Aristotle's *Physics*, 8.10, 267a12–14, writes that motion "will not be continuous in itself" although "it appears continuous to the senses." I do not think this passage, which is a discussion of the *rota Aristotelis* paradox, can be reconciled with Gassendi's other remarks on motion. Nor am I happy to simply say, along with Bloch, *La philosophie de Gassendi*, 225–7, that Gassendi's claims that motion is discontinuous are merely ad hoc solutions to Zeno's paradoxes and are not important to Gassendi's theory as a whole. Palmerino, "Galileo's and Gassendi's Solutions to the *Rota Aristotelis* Paradox," 412–13, argues against Bloch and gives an account of the historical sources for Gassendi's view of the discontinuity of motion. However, she, like Bloch, sees a closer relation between this issue and the question of whether atomic motion is "inertial" than I think is warranted.

seventeenth-century Aristotelians.[40] On this theory, the thrower's hand impresses a force or power on the stone, which force becomes the internal cause that maintains the motion. This internal cause was called by a number of names, impetus being only the best known. Eustachius – who explicitly connects this up with the maxim *everything that is moved is moved by another thing* – says that "the principle of motion" or "*virtus motrix*" of a projectile is the "*vis impressa*" impressed on it by the projector.[41] Buridan himself thought that impetus was naturally permanent and persists until it meets with resistance, but most others thought impetus died away naturally.[42] Thus, impetus theory involved two claims: first, that violent motion at uniform speed requires the constant action of some cause and, second, that no violent motion could persist indefinitely.

The claim that motion at a constant speed always requires the presence of some physical force or power was challenged in the seventeenth century, most famously by Newton's principle of rectilinear inertia but also, in different ways, by Beeckman, Galileo, Descartes, Gassendi, and others. Beeckman's journal reports that, in his 1629 meeting with Gassendi, he had told him his "opinion on motion, namely that in a vacuum all things that were once moved are always moved," and that he gave him a copy of his *Corollaria* (the theses he defended to become doctor of medicine).[43] In one of these theses, Beeckman differentiates himself clearly from impetus theorists, arguing that projectiles preserve the motion imparted by the projector rather than requiring any ongoing force.[44] It is hard to know how much influence meeting Beeckman really had on Gassendi, but the similarity to Gassendi's later view is notable.

In the Third Day of the *Two New Sciences*, Galileo's spokesman Salviati argues that when moving along horizontal planes, "the moveable is . . . indifferent to motion and to rest, and has in itself no inclination to move in any direction, nor yet any resistance to being moved."[45] Thus, a body already moving on a horizontal plane will (in idealized conditions) move perpetually with uniform motion. Galileo establishes this conclusion by reference to various inclined planes, but as Salviati's remark that "by [the horizontal] we understand a surface every point of which is

40 For an account of sixteenth-century modifications of impetus theory, see Wallace, *Prelude to Galileo*, 324 ff.

41 Eustachius, *Summa philosophiae quadripartita*, 113 (1.3.7).

42 Gabbey and Ariew, "Body: The Scholastic Background," 444.

43 Beeckman, *Journal tenu par Isaac Beeckman de 1604–1634*, 3.123.

44 Clark, "Pierre Gassendi and the Physics of Galileo," 353.

45 Galileo, *Two New Sciences*, 172.

equidistant from this same common center"[46] makes clear, uniform and perpetual motion must really be along an arc like the surface of the earth. For the uniformity of such motion derives from the fact that heavy bodies neither have a tendency to move themselves horizontally (as they do downwards) nor a concomitant resistance to being moved (as they do upwards). On Galileo's view, motion downwards needs no cause beyond the nature of heavy bodies, but uniform and perpetual motion needs no cause in the stronger sense that such motion is indifferent. This claim is most famously made in the *Letters on Sunspots*:

> [A]ll external impediments removed, a heavy body on a spherical surface concentric with the earth will be indifferent to rest and to movements toward any part of the horizon. And it will maintain itself in that state in which it has once been placed; that is, if placed in a state of rest, it will conserve that; and if placed in movement toward the west (for example), it will maintain itself in that movement.[47]

The absence of physical causes, for Galileo, leads to uniform and perpetual motion.

Descartes, by contrast, holds that uniform and perpetual motion is rectilinear rather than circular and that it is grounded not just in an absence of causes but also in the immutability of divine conservation.[48] For Descartes, motion has a twofold cause: first, "the universal and primary cause," which is "no other than God himself"; second, "the particular cause by which an individual piece of matter acquires some motion that it previously lacked".[49] Thus, particular causes are only needed for new motion and not for the continuation of motion. Descartes's first law of motion states that "each thing . . . always remains in the same state, as far as it can, and never changes except as a result of external causes."[50] This general law of persistence is spelled out in the second law: "every piece of matter, taken as an individual, does not tend to move along an oblique

[46] Ibid., 171.

[47] Galileo, *Discoveries and Opinions of Galileo*, 113.

[48] This issue has received a great deal of discussion in the secondary literature. See, for instance, Garber, *Descartes' Metaphysical Physics*, 389; and Hatfield, "Force (God) in Descartes' Physics," 121–2.

[49] Descartes, *Principles* 2.36/*Oeuvres de Descartes* 8.61. Although Descartes also refers to "certain rules or laws of nature . . . that are secondary and particular causes" of motion (*Principles* 2.37/ *Oeuvres de Descartes* 8.62), these are not independent causal agents but rather ways of talking about the divine activity.

[50] Descartes, *Principles* 2.37/ *Oeuvres de Descartes*, 8.62.

path but only in a straight line."[51] The claim that rectilinear motion persists is also justified by divine immutability, together with the premise that "everything that moves is determined, at the individual instants that can be designated as long as the motion lasts, to continue moving in the same direction according to a straight line."[52] Indeed, since we live in a plenum and hence never observe genuinely uniform and perpetual motion, it could hardly be justified by observation.

Gassendi generally agrees with Descartes that rectilinear, not circular, motion persists perpetually in the absence of intervening causes. However, Gassendi's argument relies on the assumptions about gravity and the void discussed earlier rather than on metaphysics. He also seems to assume that the persistence of uniform motion requires *some* underlying secondary cause. As we shall see, he ultimately locates that cause in the *vis motrix* of atoms – an assumption that raises a number of problems for understanding the relation between atomic and bodily motion even though it grounds a metaphysically satisfying account of the source of motion.

It must be noted that Gassendi occasionally says that *circular* motion is what persists perpetually. For instance, he claims that the motion of the celestial bodies is "uniform and perpetual, because of the circular form chosen by its author, according to which, lacking a beginning and an end, it can be uniform and perpetual" (3.488a).[53] He refers elsewhere to "not only circular motion but also its perpetual course" (1.638b) – which results from "the permanent compaction [of atoms into celestial bodies] and the texture of the globe, and hence the permanent causes of interior circular blows and circular conduct" (1.638b). This claim makes clear that Gassendi thinks uniform and perpetual motion does ultimately require a physical cause, namely, the motion of the atoms comprising the moving body. I return to this issue and its implications for the relationship between Gassendi's atomic physics and his physics of bodies later. Here I

[51] Descartes, *Principles* 2.39/ *Oeuvres de Descartes* 8.63. Descartes had accepted – perhaps, like Gassendi, as a result of discussions with Beeckman – a roughly similar view long before this. On November 13, 1629, he wrote to Mersenne that "the motion impressed on a body at one time remains in it for all time unless it is taken away by some other cause" (*Oeuvres de Descartes* 1.71). The view is well developed by *Le Monde* (*Oeuvres de Descartes* 11.38–46).

[52] Ibid., 8. 64 (*Principles* 2.39).

[53] These claims are particularly odd because, at least in the second letter *De motu*, Gassendi accepts Kepler's claim that planetary orbits are elliptical (3.515b).

just want to note that Gassendian uniform and perpetual motion is more complicated than the argument of *De motu* suggests.

Gassendi's presentation in *De motu* is centered around a defense of the "Galilean theorem"[54] that "If the body we are on is in motion, everything we do and all the things we move will actually take place, and appear to take place, as if it were at rest" (3.478a). I call this *Galilean relativity.* Gassendi begins by recounting a sailing trip that had taken place in October of 1640 under the observation of his patron Louis de Valois. On a trireme leaving Marseilles, traveling at a speed of sixteen miles per hour, Gassendi arranged for a ball to be dropped from the mast of the ship. The ball landed at the base of the mast, rather than, as Gassendi took his audience to expect, behind the mast at the point where the masthead had been when the ball was dropped. The tone of Gassendi's report suggests that the expedition's purpose was rhetorical rather than experimental: He intended to provide a public demonstration of Galilean relativity rather than a confirmation of it. However, Hobbes paraphrases their fellow Parisian and member of the Mersenne circle, Thomas White, doubting that the experiment would work as late as 1642: "[T]he instance of the stone that is released and falls close to the mast of a ship in motion White alleges to be unreliable and contrary to the belief of all those who, so far, claim to have made such an experiment."[55] Gassendi's elaborate description and noble witnesses, then, had some point within his own intellectual community as well as for outsiders.

Galilean relativity is related to uniform and perpetual motion, as follows. Gassendi ultimately wants to explain why a projectile thrown straight up from a moving vessel lands in the same spot on the deck it took off from. Because the ship has moved forward while the stone is in the air, the stone must have some horizontal motion. But what would cause this horizontal motion? It cannot be any force impressed by the thrower's hand, since *ex hypothesi* the stone was thrown directly upward. If the then standard Aristotelian account of the cause of motion were right, it would seem that Galilean relativity had to be false. Thus, Gassendi offers an alternative account of what keeps the stone in forward motion – one that

[54] What Gassendi has in mind is the claim of the Second Day of the *Dialogue Concerning the Two Chief World Systems* that "Motion, in so far as it is and acts as motion, to that extent exists relatively to things that lack it; and among things which all share equally in any motion, it does not act, and is as if it did not exist. . . . motion which is common to many moving things is idle and inconsequential to the relation of these moveables among themselves" (116). Galileo never labels any such claim a theorem.

[55] Hobbes, *Thomas White's De Mundo Examined*, 246 (21.13).

operates in terms of a uniform and perpetual motion possessed by both the ship and the stone.

He argues for uniform and perpetual motion on the basis of the same type of thought experiment he used for acceleration in free fall. Suppose a body is at rest in a large void like the extra-mundane void. The body would not start moving, for there is no reason for it to begin to move or to move in any particular direction. Gassendi takes this to be uncontroversial. The more controversial claim he adds is that a body located in a large void *within* the world would behave in exactly the same way as one in the extra-mundane void, and this addition depends upon understanding *gravia*, heaviness, as an external cause of motion.

Given this account, a body at rest in a void within the world should remain at rest. Gassendi's next step is to ask what would happen if that body were put into motion:

> [Y]ou ask, what will happen to the stone that we assumed could be conceived in those void spaces, if it were disturbed from rest and impelled by some power [*vis*]? I respond that it is probable that it would move uniformly and indefinitely, and indeed slowly or quickly just as a small or a great impetus were impressed on it. (3.495b)

He goes on to add that

> motion impressed through void space, where nothing either attracts or restrains and nothing at all resists, will be uniform and perpetual. And from this, we conclude that all motion impressed on a stone is in fact of this kind. And thus, in whatever direction you might throw a stone, if you suppose that at the moment it is let go by the hand everything except the stone is reduced to nothing by divine power, it will happen that the stone will move perpetually and in the same direction in which the hand has directed it to move. (3.496a–b)

For, despite his use of the term "impetus," Gassendi does not think that putting a body into motion requires anything to be transmitted from hand to stone:

> [I]t is clear that nothing is impressed [on the body] by its mover other than motion. . . . Such motion as the mover has while it is conjoined with the moved thing is impressed [on the moved thing] and will continue the same and be perpetual unless it is destroyed by some opposed motion. (1.499a)

The claim that nothing save motion (that is, no force or impetus) is impressed on the stone could easily be read as the claim that uniform and perpetual motion requires no cause. Indeed, so far as the account of *De motu* goes, this seems to be pretty much correct. However, when we turn to the question of how Gassendi's atomic and macroscopic physics of motion interact, we shall see that the situation looks quite different.

Atomic Motion and the Motion of Composite Bodies

Like many of his Parisian contemporaries, Gassendi wanted to provide a properly causal, natural philosophical basis for Galileo's claims about motion. His attempt to do so involves both a corpuscularian mechanism, the chains of gravity connecting falling bodies to the earth, and at least the suggestion that uniform and perpetual motion relies on the intrinsic *vis motrix* of atoms. Thus, one can see Gassendi as trying to ground theories of motion in matter theory. Moreover, as we saw in the previous chapter, Gassendi makes it very clear that the ultimate cause of the motion of composite bodies is atomic motion. Consider the following passage, one of his more detailed explications of this causal relationship:

> [T]here seems to exist in [natural things or composite bodies] an innate motion on account of the atoms. And either these atoms are those from which the mobile body is composed, as when it is moved by itself or *per se*, or they are those that make up the moving body, as when something is moved by something else, which thing, while it does the moving, would in some measure be moved by itself. For, since the atoms in some body are variously agitated, if some that are more mobile and quicker [*expeditiores*] conspire together to press towards some place, then the whole body is itself moved in that direction. (1.338a)

It is difficult to make out exactly what the relation between Gassendi's atomist matter theory and the accounts of motion put forward in *De motu* and *De proportione* is. At the atomic level, matter is described as being active: Each atom has an innate *vis motrix*, a power created in it by God that requires nothing beyond the concurrence required by everything in nature to continue acting. This *vis motrix* is responsible for the perpetual motion of atoms. However, at the level of composite bodies, motion is uniform and perpetual so that no new cause is required to keep a body in motion once it has begun to move. Once a body begins moving it will continue to move at the same speed and in the same direction, until it is affected by the impact of atoms, corpuscles, or concretions from outside. In fact, the uniform and perpetual motion of composite bodies must ultimately be explicable in terms of the *vis motrix* of each component atom. Thus, uniform and perpetual motion results when atoms with unchanging weight are combined so that they keep moving in the same direction, and accelerated or decelerated motion, or change in direction, results from an external blow changing the pattern of vibration comprising a compound body.

All the phenomena of motion of composite bodies ultimately have as their physical causes the motion of atoms, on Gassendi's view. However, he has no real account of how certain patterns of atomic motion yield macroscopic motion. The looseness of Gassendi's treatment of the two essential notions, texture and vibration, becomes particularly apparent here. The texture of a body depends on the pattern of vibration of the atoms making it up, as well as on the intrinsic qualities of the atoms themselves. Such textures are supposed to fill a number of roles. They are supposed to ground the sensible and occult qualities of bodies. Their component patterns of vibration are supposed to explain the speed at which composite bodies move, at least on the assumption that individual atoms are always moving at maximum speed (1.343b). It seems likely that patterns of vibration are also supposed to determine why a composite body is moving in the direction it is. However, at no point does Gassendi attempt to work this out even at the most general level. Nor is it clear how he could do so. The Galilean treatment of falling bodies, for instance, assumes that textural differences between bodies are irrelevant – the proportion of acceleration is not dependent on specific bodily characteristics. But why, on Gassendi's view, shouldn't it be relevant? Why should we expect the rays of corpuscles emitted from the earth to act on bodies of all textures, composed of atoms of all different sizes and shapes, in exactly the same way? Gassendi assumes that they do – and he must assume this, for (virtually) all sensible bodies are affected by gravity. However, he has no principled account of why this is so and no other evidence for thinking that the "hooks" of gravitational corpuscles grapple with all types of bodies equally well.

Does Gassendi fail to provide the sort of explanation he owes his readers, or would he not think he owes us any such explanation? Three different considerations should be borne in mind. First, Gassendi makes it clear that the truth of atomism is supposed to follow from the mere fact of motion: We do not need any particular data about how motion occurs to know that the world is composed of atoms in the void. Thus, it is open to Gassendi to remain silent about which particular version of atomism is the correct one. He need not be able to provide a complete theory of how atomic motions combine in order to claim to know that atomism is true, given his methodology. And in fact, one suspects, he would take it to be unconscionably dogmatic to do so. Thus, it is open to Gassendi to insist that there need be no clear connection between atomism and the account of composite motion: All that is needed is that it is *possible* for bodies composed of atoms to combine to form bodies moving as has been observed. And although he provides no derivation of composite motion from atomic motion, he gives no reason to think that such a derivation is in principle impossible.

Second, the gap between Gassendi's atomic and composite accounts of motion helps make clear how little connection there really is between the program sketched in the opening book of the *Syntagma*, "De rebus naturae universe," and the more empirical aspects of his natural philosophy. Atomism rules out some possibilities for explanations, but it does not really act as a *source* of natural philosophical explanations in most cases, and certainly not in the case of the motion of compound bodies. Just as Gassendi's stated methodology makes it permissible for him not to fill in the gap between accounts of atomic and compound motion, it also makes it permissible for him to give explanations of particular phenomena piece-meal.

Finally, the gap between Gassendi's two theories of motion helps point out the great importance the goal of preserving secondary causation apparently had for him. Gassendi's suggestion that atoms are intrinsically moving, because they are possessed of an intrinsic motive power, grounds the reality of secondary causation. And it does so in a way that accounts of motion that took it to belong to matter without being intrinsic to matter could not do. Here we are far from a picture of Gassendi the empiricist, embracing atomism for its practical explanatory value. Rather, we see Gassendi addressing what is for us a paradigmatically metaphysical issue that arose out of a dissatisfaction with both Aristotelian and neo-Platonist accounts and coming up with a theory that has virtually nothing to do with observation and everything to do with the requirements of his moral and religious world view.

8

Generation, Life, and the Corporeal Soul

Along with the vast majority of his philosophical tradition, and in contrast with Descartes, Gassendi distinguishes between inanimate bodies on the one hand and plants, animals, and human beings on the other.[1] The process of generation crucially requires powers that cannot be reduced to the powers of atoms taken singly or woven together into concretions. These powers are either emergent or divinely super-added to certain systems of matter; Gassendi, as we shall see, offers two alternate cosmologies grounding each of the two possibilities. Thus, the programmatic atomist claims that Gassendi advances at the beginning of the *Syntagma* are implicitly restricted to the nonliving. He does indeed hold that there are no forms or qualities beyond ways of describing complex local motions. But he does not think of generation as a matter of qualitative change any more than Aristotle did, and hence generation is not intended to fall under the scope of claims about the reduction of qualities to local motion.[2]

Similarly, the sentient powers of animals are either emergent or super-added. Indeed, they derive from the same seeds that are crucial for generation, seeds endowed with a sensory capacity and a *vis seminalis.* Animal souls have powers beyond sensation and generation, of course: In addition to providing the power of locomotion, they contribute powers shared

[1] I do not say that Gassendi sharply distinguishes the living from the nonliving, for he classifies only animals possessing vital heat as living, excluding plants. Although I do not know how common this practice was, it can at least be found in Fernel, *Physiologia,* 260–5 (Section 4.2).

[2] Aristotle, *De generatione et corruptione,* II 319b6–7.

by plants and animals such as growth. However, I limit my discussion to two powers, generation and sensation.

The soul I discuss here is the corporeal soul possessed by plants and animals as well as human beings. Following Lucretius, Gassendi calls it the *anima*, as opposed to the *animus* or incorporeal rational soul unique to humans. I discuss the incorporeal soul and its relationship with the corporeal soul and the body more generally in Chapter 10.

The Powers of Souls

A corpuscularian account of the soul sounds radical within Gassendi's historical context. However, his materialist theory is rather more conservative than one might expect. For one thing, Gassendi ends up identifying the soul with the principle of life and hence with vital heat, and such identification is entirely traditional. Fernel's influential *Physiologia*, for instance, claims that "all philosophers are of one mind in reckoning life by heat" because heat is commonly understood to be both a *sign* of life and a *cause* of life.[3] Gassendi holds that the vital heat of the soul is a product of corpuscular or atomic action, but he has no developed account of what this action consists in. His treatment of heat, and hence of a number of essential functions of the soul, is structurally very similar to his treatment of the chemical elements. As we saw in Chapter 2, Gassendi changes the ontological status of the elements drastically while assigning them more or less the same causal powers and role in explanation as his predecessors. The same could be said of his account of the powers of the soul, which becomes a purely material body while retaining the same causal roles.

What are these causal roles? In his *De Anima* commentary, Aquinas ascribes three powers to the vegetative soul – nutrition, growth, and generation – and five more to the sensory soul – sensation, passion, memory, imagination, and locomotion. In any given plant or animal, these powers all come from the same unitary source: Animals do not possess both a sensitive and a vegetative soul, but rather one soul that has all eight powers.[4] In contrast, Galen delineates a clear separation between the parts of the soul, localizing the vegetable soul in the liver (where it operates through the veins), the natural or vital powers of the animal soul in the heart (where they operate through the arteries), and its sensitive

[3] Fernel, *Physiologia*, 259 (Section 4.2).

[4] See Aquinas, *A Commentary on Aristotle's De Anima*, 225.

power in the brain (where it operates through the nerves). Thus, he adds life, as a distinct power from nutrition or growth, to the Aristotelian list.[5]

This tripartite division of the animal soul was held to have implications for embryological development, the most straightforward of which is that the brain, heart, and liver must develop first. Gassendi's insistence that the powers of the soul are inseparable is tied in with his anti-Galenist account of the order in which the organs are formed. We shall see that his embryology places a great deal of emphasis on this point. However, he is perfectly willing to accept talk of various parts or faculties of the soul, and indeed provides a complete exposition of each power in typical textbook order.

While life, growth, nutrition, sensation, passion, and locomotion are straightforwardly powers of the soul on Gassendi's account, generation has a somewhat more complicated status. For although generation is closely allied with growth and nutrition, properly speaking generative powers inhere in seeds or *semina rerum* rather than a soul or an entire plant or animal body. This point is helpful in seeing how Gassendi can maintain that generation, or a process very much like it, forms inanimate crystals and minerals.

Gassendi's interest in the generation of crystals began during his early years working with Peiresc, whose mineralogical views he describes in his biography:

> [Peiresc] was not of the opinion, that all stones were created at the beginning of the World: but he conceived, that in progress of time many were made in such manner, as to owe their original to their matter, and certain peculiar seeds; receiving their shape, partly from nature, partly by chance. For the matter of all them being water, or some other liquor or juyce, he conceived that in divers places were contained divers seeds of things, and particularly of stones, which being mingled with the liquor, does curdle the same, as milk is curdled by the runnet, and imprint its particular form thereupon. Consequently, that Crystals, Diamonds, and the like stones are made, when their seeds meet with a transparent liquor, such as they are only capable to perfect, and other stones, when their seeds meet with a more troubled and obscure Liquor . . . while they are in coagulation, they are parted, and multiplyed, as the grains of an eare of corn, within the sheath . . . sprung from their peculiar seeds; by means of which, stones receive their proper shapes as constantly and regularly, as Plants and Animals. (*Mirrour* 4.46–7)

[5] For the separation of the soul into three parts and the localization of the parts of the soul, see Galen, *On the Doctrines of Hippocrates and Plato,* 1.373 (VI.3).

Gassendi adds that he and Peiresc took gemstones to be composed of salt and alum mixed in various proportions, a claim whose chemical heritage, like that of Peiresc's own view, is unmistakable. Thus, one might suspect that it is a chemical notion of generative power that pushes Gassendi toward separating the generative powers of *semina* from the powers of the soul.

Gassendi explains the generation of plants and animals from seeds in the same language he uses to explain crystal formation. He also introduces further chemical terminology: In language perhaps deriving from Severinus, he explains *semina* as corporeal entities endowed by God with powers of *scientia* and *industria* that shape and determine the offspring they produce. He appeals to such *scientia* and *industria* to explain those elements of generation, growth, and nutrition that appear goal-directed. Thus, he reinterprets the apparent finality of generation by reconstructing the goal of generation – the mature plant or animal – as contained within the seed's idea of the mature offspring. Gassendi understands the idea within the seed as a sort of miniature, material representation of the offspring, and thus the suggestion has certain preformationist elements. Spelling this out will require a closer look at the notion of *semina rerum*.

The Powers of Seeds

The concept of *semina rerum* has a long history within a number of traditions, and the complex explanatory role *semina* play within Gassendi's physics is a prime example of his syncretism.[6] Visible seeds like the seed of an apple are, of course, *semina*, but so are many other things. Gassendi's *semina* are corpuscles or molecules – insensibly small concretions of atoms whose stability and coherence allows them to play a role in explanation – and sometimes he speaks as though *all* corpuscles are *semina*. For instance, he explains that "there are not only Atoms, but also seeds or molecules, which are themselves . . . far beneath the senses" (1.493b). More often, however, he uses the term *semina* for the molecules specifically relevant to the generation of plants, animals, and minerals.

Gassendi's interest in the generation of minerals predated his writings on plant and animal generation considerably. Finding that stones, salt, snowflakes, gems, dust particles, and even bladder gravel all had regular

[6] For previous discussions of the role of *semina* in Gassendi's thought, see Bloch, *La philosophie de Gassendi*, 252 ff; Clericuzio, "Gassendi, Charleton and Boyle," 476–81; and Osler, "How Mechanical Was the Mechanical Philosophy?" 437–8.

geometrical shapes, Gassendi arrived at the conclusion that this regularity could not be accidental, but rather happens "because of a constant cause, a sort of seminal virtue" (2.114 ff). This *vis* or *virtus seminalis* is introduced to explain the regularity and structure of stones as well as that of plants and animals. He treats the question why dogs always give birth to dogs rather than, say, lizards, in the same way as the question why crystals of the same type always have the same geometrical shape, and this tendency to treat mineral and animal generation in parallel gives a certain push toward preformationism. Hexagonal crystals are subdivisible into the hexagonal seeds or atoms that formed them, so that the seed is a microcosm of the whole. Similarly, the seeds contained within animal bodies contain a little version of the adult animal, and the soul within the seed contains a rudimentary version of the adult's soul. Given Gassendi's rejection of formal causality distinct from matter, he is inclined to reject generation *de novo*, whether of plants or of animals, because without some structure in the initial matter he sees no way for structure to develop. *Semina* are supposed to explain what this preexisting structure consists in.

It may be useful to look briefly at the background history of *semina rerum* before we consider Gassendi's own account. In Lucretius, *semina rerum* are atoms.[7] *Semina* or *semina rerum* also play a role for neo-Platonists like Ficino and, following Paracelsus, for many chemists. Consider, for instance, Jean-Baptiste van Helmont, Étienne de Clave, and Petrus Severinus, all of whose work Gassendi had some familiarity with.[8] Paracelsan *semina* have both spiritual and corporeal aspects, which makes it somewhat hard to see how they could fit into an atomist worldview. But later developments place less stress on the incorporeal, spiritual elements of *semina*. Nicholas Hill's *Philosophia Epicurea*, for instance, refers *semina* to the Paracelsan tradition and defines them as "certain atoms brought together."[9]

Severinus is particularly interesting here. For him, *semina*, *astri*, or *radices* contain "the Mechanical arts and dispositions" of nature and "all

7 Lucretius, *De rerum natura*, 1.58–61. Lucretius also uses the term *semina* (though not *semina rerum*) as a synonym for *concilia*, where *concilia* are the small compound bodies associated with the four elements (1.741–6). However, this usage is infrequent and not particular to generation; Gassendi is not deriving his use of the term from Lucretius.

8 Gassendi corresponded with van Helmont, although as far as I know only on the subject of the relative merits of vegetarian and carnivorous diets (6.19b). He refers to de Clave and Severinus in the *Syntagma* (see, for example, 2.554b). Bloch, *La philosophie de Gassendi*, 446 ff, gives some additional manuscript evidence of familiarity with and interest in Severinus.

9 Hill, *Philosophia Epicurea, Democritea, Theophrastica*, 8.

vigor and power is derived" from them.[10] He also speaks of *semina* as possessing *scientia* and *industria*, features crucial in Gassendi's account of generation. For Severinus, the *scientia* that seeds possess amounts to something like a plan for the construction of bodies from undifferentiated matter. Things "flow forth" or "progress" from *semina*; at the same time, Severinus explains that seed-governed development is a *mechanicus processus* guided by *scientia mechanica*. It is helpful to keep this language in mind when reading Gassendi's description of seeds as *machinula*, little machines that turn out to contain other, smaller machines within them (2.267a). For the powers that Gassendi and Severinus ascribe to seeds are certainly not mechanical in any of the senses of the term used by recent scholars. A *mechanicus processus* is simply a craftsmanlike process.[11]

I do not intend to suggest that the mere fact that Gassendi uses the term *semina* is evidence of any Severinan or, indeed, chemical background. The notion of seeds as primary in generation was of course widespread, given that *semina* are in the first place ordinary, visible, agricultural seeds. Moreover, the language of seminal powers can be found in Aristotelian textbooks and even in Descartes. In the *Discourse*, Descartes explains that when he wrote *L'Homme* he could not explain human and animal bodies in the same way he explained the phenomena of *Le Monde*, namely "by demonstrating effects from causes, and showing from what seeds, and in what manner, Nature must produce them."[12] However, the claim that seeds possess *scientia* and *industria* is far less common, and it is this claim that suggests a connection between Gassendi and Severinus.

Gassendi's *semina* are corpuscles or textures of atoms, and differences in the textures of seeds play a basic explanatory role in his theory. Different seeds act differently "[b]ecause the interior texture of all *semina* is not of the same kind, and thus . . . those branching-points [*divaricationes*] through which the soul is collected, strives and is determined are of one sort in one seed and of another sort in another seed" (2.262a).[13] This language suggests that Gassendi will describe how seeds act in terms of their texture. However, when he treats their particular actions he tends to switch from the language of textures and atoms to the language of

[10] Severinus, *Idea Medicinae Philosophicae*, 6.49.

[11] Shackelford, "Seeds with a Mechanical Purpose," 231.

[12] Descartes, *Oeuvres de Descartes*, 6.45.

[13] I have benefited greatly from the translation of portions of Gassendi's "De generatione plantis" and "De generatione animalium" in Appendix A of Adelmann, *Marcello Malpighi and the Evolution of Embryology*.

chemistry. We should consider whether this switch is substantive or merely a matter of convenience.

Gassendi provides a way of understanding the ontology of chemistry in atomistic terms: The chemical elements are actually certain types of corpuscles, "proximate and immediate principles" of bodies (1.472a). Is this a reductionist view? It is easy for contemporary readers to take it as such, but we should be careful about doing so. Gassendi's insistence that corpuscles, and indeed all macroscopic bodies, are no more than atoms arranged in the void by no means commits him to a denial of emergent or superadded powers.

It turns out that Gassendi's account of the generative powers of seeds has clear antireductionist suggestions, and I devote a large portion of this chapter to reconstructing this account and its ontological significance. Two questions are crucial: What is the role played by the *scientia* and *industria* of seeds? And what precisely is the involvement of God in the generative process? These questions are, of course, closely connected: God endows seeds with special powers, and our awareness of these powers provides a good way for us to come to recognize the wisdom and power with which he designed the world. However, Gassendi combines the claim that generation manifests God's direct involvement in the world with an entirely naturalistic account of generation. I begin by looking briefly at the generation of plants, and then in rather more detail at the generation of animals. Starting with plant generation will help us see why Gassendi thinks the structure of adult organisms must be present in the seed, despite the difficulties this creates for explaining how both parents contribute to the new animal.

The Generation of Plants

Gassendi's account of generation has two central ingredients: the seed and the soul, or principle of vegetation. He explains that

> [B]ecause the soul of the plant is corporeal, it can only be understood as a certain substance diffused through the plant, a substance like a spirit or little flame, which is singularly fine, pure, active, and industrious, and which withers on account of deficient aliment, is suffocated because of too much humor, or is evaporated because of heat. (2.172a)

Plant souls contain within themselves a representation of the entire plant:

> The soul of the plant is a certain highly spiritual and active substance, all of whose parts communicate among each other so that in whatever part of the plant it is,

> [the soul] contains as it were the idea and impression of the other parts. Thus, it has [the idea and impression of all the parts] most powerfully in the seed, toward which everything tends from the beginning and in which [everything] ultimately terminates. (2.172b)

Both the soul as a whole, then, and the soul of the seed contain ideas of all the parts of the future plant. Seeds and their souls are epitomes, little miniatures, of fully grown plants and their souls. We shall see how this works in more detail in the case of animal generation, but here it is sufficient to note that a seed's idea and impression is both the idea and impression of the plant that produced it and the idea and impression of the plant that it will grow into. Gassendi is committed to the view that each plant is a copy of its parent; no difference, save perhaps a failure of perfect reproduction, is possible.

Just as adult plants have souls, so seeds have little souls, *animulae.* The most important feature of the *animula* is its *vis seminalis*, and it is in virtue of that seminal power that the *animula* can guide the development of the epitome it contains into a mature plant:

> The *animula* enclosed [in the seed] begins separating all these particles [of seed] and as it were distributing their regions and governing the work of each. And also the particles, by means of some kind of will, remove themselves from the confusion and similar [particles] approach each other and unite. (2.173a)

The language Gassendi uses is more intentional than it needs to be. For instance, although his claim here is that like particles approach each other by will, the way he elaborates the account makes clear that the mutual attraction of similar particles is supposed, in some rather mysterious and ill-explicated way, to result from their *vis motrix* (2.266a).

Gassendi's account of the way new matter is acquired and distributed suggests the sort of attempt to assimilate generation to growth manifest in later, preformationist theories. For instance, he argues that the entire structure of the adult plant is contained within the bud and simply needs to increase in size:

> When the bud sprouts, it is apparent that the, as it were, foundation-threads [*stamina*] of the whole shoot have already been formed in [the bud]. Because they are subtle and pressed together in a mass, they only need to unfold and increase in order to be more clearly seen. Hence upon inspection the texture of the shoot can be recognized, with its pith, wood, bark, and also its little leaves, and in the places where the branches connect to the stem indication of the buds to be produced in this or a following year. And if a flower needs to emerge when the branches are adult, the rudiments not only of that flower but also its fruit can be recognized. (2.185b)

The claim that even indications of future buds are present in a newly formed bud makes clear that Gassendi is not only rejecting generation *de novo* but also assuming that no new structure comes into being at the moment of germination. Instead, germination merely allows the structure already there to develop. The presence of warmth and moisture cannot cause the emergence of a new, structured organism, although it is a necessary condition for that structure to unfold itself. Gassendi's insistence that "spontaneous generation" is simply generation from a seed randomly assembled in the earth derives from this way of thinking about generation: If no entirely new structure can come into being, then no structured organism could come into being from an undifferentiated mass of mud.[14]

How does the claim that the structure of future plants is contained within those now alive relate to the central role played by the *scientia* and *industria* of seeds? The two kinds of terminology tend not to be combined in the same passages. In some sense, the preformationist element and the *scientia* of seeds both fill the same role, namely the role of explaining how apparently unstructured matter could come to possess structure. However, preformed structure and *scientia* operate in sequence. The preformationist element explains the structure present at the moment of germination. This is exemplified in Gassendi's various claims that the real problem of generation is the problem of the generation of *seeds*:

> Generation is ordinarily understood to be the actual fashioning of the members from the *semen*, but we can say that a thing is really first generated at the time when the *semen* from which it comes is procreated; for of course the *semen* contains the thing itself, but contains it not unfolded, and in the form of rudiments. (2.262a)

The *scientia* of seeds, in turn, explains how the epitome or preformed structure grows into a mature organism. For instance, *scientia* is responsible for the right sort of matter being added in the right place.

Gassendi argues for this view mainly on a priori grounds. As we have seen, he does claim that a structured miniature can be observed in the bud, and he makes similar claims concerning chicken embryos in eggs.

[14] Gassendi compares the generation of animals through copulation with allegedly spontaneous generation: "the former are not produced or born except from animals very like them, or animals of their own kind; the latter burst forth and spring up even from dissimilar things . . . the latter just as the former owe their origin to their seeds, but the former are customarily said to have their origin from seeds, the latter from putrid or other matter, insofar as the seed of the former is manifest within a similar animal and the seed of the latter is hidden in alien and as it were unlooked-for material" (2.263a).

However, I do not find it plausible that these observations are supposed to bear a great deal of weight. Rather, Gassendi's account of generation is constrained by several previously developed theses about causation. The account cannot appeal to forms, and hence not to formal causation. Nor, as we shall see in more detail, can it use natural teleology. God's intentions are obviously relevant, but a complete story would provide the secondary causal mechanisms by which those intentions are carried out: The account must operate in terms of secondary causation within nature. These constraints, taken together, render impossible the Aristotelian and Galenist claim that structure develops out of undifferentiated matter and thus lead Gassendi to posit that structure has always been present.

It will be useful to compare Gassendi's account to that of another atomist like Daniel Sennert, which has been helpfully explicated by Emily Michael.[15] Michael locates Sennert as a follower of the minority view on which seeds transmit a complete specific form or soul at the moment of conception, so that a lion and the seed of a lion have the same soul. This view is in competition with the older Aristotelian view, still the view of the majority, that the soul or specific form of a new entity does not come into being until after the seminal matter has been shaped. Sennert favors the minority view because of his answers to two questions that structure his account. First, what causes a particular bodily structure to develop in embryological generation? Second, what is the cause of a soul?

Sennert insists that embryological generation requires some specific, prior, organizing principle: the soul or (what is in this context the same) the plastic principle. Since growth begins at conception and the soul or specific form is the principle of growth, the soul must be present from the beginning. In fact, the soul results from the combination of seeds from the two parents and comes into existence at the very moment when the seeds are combined. Michael emphasizes a point that is important for our purposes: Namely, that for Sennert the form is the *efficient* cause of growth and development, although in more traditional Aristotelian accounts it could not be an efficient cause because it did not come into being until the organism was completely formed. Gassendi is not willing to call the idea in the seed a form, but structurally his account is very similar. The idea of the mature organism that is contained in the soul of the seed is the efficient cause of generation and thus provides a nonteleological explanation of the apparently goal-directed processes involved.

[15] Michael, "Daniel Sennert on Matter and Form."

The Generation of Animals

Gassendi's account of animal generation is structurally very similar to his account of plant generation – indeed, more similar than is really plausible, given how serious a problem it is to explain how the seeds from both parents combine to produce an embryo. Like his contemporaries, Gassendi follows the Galenist two-seed theory of generation on which both parents contribute structure and matter to the new animal. Aristotle's view that the female parent contributes only matter was still known but no longer taken seriously. Jean Fernel, a relatively conservative sixteenth-century figure, wrote that "[i]t seems fanciful and very like an old wives' tale [*anicularum figmentis*] that the male's semen alone begets something complete out of the maternal blood without any assistance from the woman."[16] Similarly, Gassendi insists that "it is from both parents that . . . the first *stamina* are formed" (2.273b). When the seeds of the two parents come together, they must "subject themselves to one another so that a single something results from the mixture of the two" (2.273b). Thus, a central problem for Gassendi and his contemporaries is to explain how offspring can inherit traits from both parents.

Gassendi's account of generation can be seen as an attempt to answer three main questions:

(1) What is the source of generation? That is, where does the new animal come from? What is the first thing generated, and what generates it?

(2) How does this starting-point develop into an animal? That is, what is the sequence of steps in embryological development, and what guides those steps?

(3) Why do these steps produce something like the parents? (One could, in fact, distinguish two different versions of this question. Why do these steps produce something the same species as the parents? And why do they produce something that particularly resembles the individual parents? However, Gassendi, who draws species boundaries just in terms of primitive similarity, runs the two together.[17])

I discuss how Gassendi answers the three questions in turn. It will be apparent that he devotes far more time to the second. His answers to the

[16] Fernel, *Physiologia*, 559 (Section 7.7).

[17] Ibid., 591 (Section 7.12), actually distinguishes three questions, having to do with "likeness in form, likeness in sex, and likeness in outward appearance."

first and third questions, which require an explanation of how the seeds from both parents are combined into a structured individual, are rather more tentative.

Gassendi begins by explaining that the primary difficulty concerns "the formative faculty and the way in which it acts" and in particular how it produces an animal like the parents "from an amorphous humor" (2.274a). But his ultimate answer, here as in the case of plant generation, is that no new organism *is* produced from an amorphous humor. What the mixing of seeds produces already contains the rudiments of the mature animal and a soul that uses its *scientia* and *industria* on these rudiments to help them grow into a mature animal.

The *scientia* and *industria* of seeds serve simply as place-holders. Gassendi insists that "man is not so fortunate in his genius" (2.273b) as to fully understand the process of development. Thus, references to *scientia* and *industria* may not be intended to remain part of a complete explanation of generation.

Gassendi is on firmer ground delineating the sequence of steps in embryological development. The first stage is the moment of conception, directly after the two seeds have combined, and on Gassendi's account all the parts of the fetus are present from that moment on:

> It seems that the formation of the membranes enveloping [the fetus], of the umbilical vessels and of all parts of the fetus begins at the same time. Thus from the very start the rough *stamina* of everything are there together, even if it does not happen that all the parts are perfected together or at the same time but some appear larger and more distinct earlier, others later. (2.277a)

Thus, Gassendi does not explain how undifferentiated matter can give rise to the complex structure of mature plants and animals. Rather, he rejects the premise that their matter ever *was* undifferentiated.

Gassendi provides a number of reasons for thinking that all parts of the animal, including the soul, are present in rudimentary form from the moment of conception on. For one thing, the simultaneous formation of the parts of the fetus is supposed to result from the superiority of natural workmanship to artificial workmanship, or divine workmanship to human workmanship:

> The works of nature do not conduct themselves in the same way as the works of art, which are only made by progressing from one part to another, since, unlike an artificer, nature together with its instruments is itself present among the matter and can act on all the parts at the same time. (2.277b)

This yields a sharp difference in kind between artificial and natural creation – a difference that Gassendi will come back to repeatedly in discussion of the divine.

On rather a different note, Gassendi argues that the interconnectedness of the parts of animals requires them all to be formed at once: "because of the very manifold, for the most part reciprocal, penetration, separation, insertion, and other things of the sort" we cannot form one part without another (2.277b). For instance, the brain cannot be formed without the heart and liver also being formed, since there are so many veins and arteries connecting the two. For the same reason, we must doubt Galen's claim that the liver is formed before the heart, and Aristotle's claim that the heart is formed first. Gassendi does not mention anything beyond the Galenic triad in this passage, but because he thinks that *all* the organs are given at least rudimentary form at once, he at least ought to say the same about the others.

Gassendi supports this rather a priori claim from considerations of the order in which the parts of the fetus are observed to develop. For instance, some claim that we can observe a fetus with a heart but no other parts. This would support the Aristotelian claim that the heart is the first thing to be formed. Others – and here I take it that Gassendi is referring to Fabricius or his followers[18] – claim to observe a stage at which fetal chicks display a head and sternum with no other parts visible. Gassendi cannot accept this claim:

> [Regarding the claim] that chicks around the fourth day [after fertilization] already show a head and a sternal ridge when no wings or legs are visible: Nevertheless the beginnings of the wings and legs are there, with all their articulations, even if they are very small, and they are immediately going to be extended. (2.277b)

Similarly, he opposes Fernel's claim that three bubbles, the heart, liver, and brain, can be seen at a very early stage of development when nothing else is there,[19] saying that what is observed "argues rather for simultaneous fashioning because there is as yet no very clear discrimination of the parts" (2.277b).

Here Gassendi is not so much claiming that fetal observations *show* that all parts come into existence at the same time as he is denying that observation has established that the liver–heart–brain triad (or some other

[18] We know from *Mirrour* 4.28–9 that Gassendi was greatly impressed by Fabricius and knew at least his work on the heart in detail.

[19] Fernel, *Physiologia*, 575 (Section 7.9).

privileged structure) develops first. Thus, observed differences must be differences in the speed at which organs grow and are perfected. Although Gassendi is clearly interested in the details of embryo development for their own sake, he is also concerned to defend the claim that all parts begin at the same time for theoretical reasons. The addition of the claim that different parts are perfected at different speeds allows him to reconcile the observations he cites with his opposition to *de novo* generation.

Let us turn to the final question, what ensures that the product of generation will resemble the parents, both in species and in individual traits. The Galenist view is that there is a "contest of seeds" in conception so that the offspring resembles whichever seed was more powerful. Gassendi attacks this with the standard objection that if the contest of seeds theory is right, male children should always resemble their fathers and female their mothers, which we know is not the case. His alternative proposal involves the objects toward which the parents' imaginations were directed at the moment of conception. For instance, he says, it is possible for a fetus

> [to resemble] neither [parent] in any way, if both their imaginations had been distracted... and in the case of the male did not have for its object the female herself, and in the case of the female [did not have] the male himself. Indeed, it seems that this can cause the resemblance that sometimes occurs not only to a statue or image but also to another man or another female. (2.284b–5a)

Gassendi recounts several famous cases in which this is thought to have happened, but rather sweetly thinks it is exceedingly rare (*Mirrour* 4.86–7). In normal cases, the imagination of each parent is devoted to the other at the moment of conception, so that the child resembles whichever parent's imagination was directed more weakly.

This sort of suggestion has a long history. More or less the same theory, for instance, was given by Fernel, who writes that contest of seed theories cannot explain why "a fetus frequently displays the form and likeness of another unknown person who has contributed nothing to procreation."[20] Fernel thinks this happens when "the power that shapes the fetus is guided and regulated by the strong grasp and steady thinking of pregnant women," as when white parents have a black child because the pregnant women was obsessed with a picture of an Ethiopian.[21] Like Gassendi, Fernel takes it that the father's imagination can also be

[20] Ibid., 591 (Section 7.12).
[21] Ibid., 593 (Section 7.12).

culpable, but, unlike Gassendi, he does not think of this as the main explanation of inheritance. Indeed, Saul Fisher has argued that the chief contribution of Gassendi's account of generation is the way his theory of maternal and paternal impressions provides a "materialist mechanism" for trait inheritance, although he grants that the mechanism is "imperfect, not always or exclusively deployed in generation, and not clearly the bearer of information in materialist form."[22]

It is difficult to tell how much faith Gassendi has in his suggestion that maternal and paternal impressions explain trait inheritance. For on almost all points, save the description of the steps of fetal development, Gassendi insists that a full understanding of the process of generation is beyond us. While discussing the formative faculty, for instance, he makes clear that the *scientia* and *industria* he ascribes to seeds are the manifestations of God's workings within nature:

> We can understand how it happens that the *animula* contained in the seed begins, advances, and perfects the elaboration of its organs and of its whole body with so much skill and *industria*, since it is a workman [*artifex*] with so much greatness, so much wisdom, so much power, who has established [the *animula*] as it is and endowed it with such force and willed it to be embraced in such a body, of such texture, that it could not act in any other way and could not but build such a structure. (2.262b)

This is not merely an admission of unhappiness with the account, if it is that at all. Rather, it emphasizes another way in which contemplating nature leads to knowledge and worship of the God who made nature: "It is proper to admire and commend entirely that divine workman, by whom alone this sort of agent can have so perspicuous an understanding of the work and so powerful and subtle a faculty of working" (2.274a). Seeds and their souls are secondary causes in virtue of their *scientia* and seminal power, but all we can say about the source of that power is that it ultimately derives from God. We are incapable of understanding the way in which these worldly powers operate, and hence we should turn our attention toward the divine. The limits of our knowledge of generation are reinforced in Gassendi's discussion of the source of the first seeds.

Divine Creation, Lucretian Evolution, and the Generation of Seeds

A seed that comes into existence now is formed from a previous seed, just as the animal or plant containing it is. But where did the first seeds

[22] Fisher, "Gassendi's Atomist Account of Generation," 2.

come from? This, for Gassendi, is the crucial question of generation, and he provides two possible answers (2.262a). One possibility is that the first seeds were directly created by God and distributed throughout the earth. On this view, the divine creation of atoms brought them into being already formed into stable concretions possessing various superadded powers. For, Gassendi writes, if God directly creates seeds and their powers, it follows that "the *vis seminalis* that he endowed the earth with at that time has not been exhausted or worn out but even now flourishes constantly, unconsumed" (2.170b). I call this possibility "direct divine creation" of seeds.

Gassendi also suggests another account of the genesis of seeds, a modified version of Lucretian evolution. In Lucretian cosmology, the world began as a random arrangement of separated atoms that came together by chance to form seeds and bodies (1.485b ff). Gassendi's version eliminates the element of chance in favor of divine creation of the original atoms and their motion. Thus, God originally created atoms that were not yet structured into any stable, coherent composite bodies. Rather, composite bodies "evolved" or unfolded from the initial arrangement of atoms without any direct divine intervention: "While [the atoms] are moving in various ways and meeting, interweaving, intermingling, unrolling, uniting, and being fitted together, molecules or small structures similar to molecules are created, from which the actual seeds are constructed and fashioned" (2.171a). Gassendi initially provides these two possibilities for the generation of seeds in the context of plant generation, but they recur in connection with the generation of animals. In that context, he seems to favor direct divine creation, suggesting that it is implied by the Biblical claim that "God created everything at the same time" (2.262b). However, his adherence to direct divine creation is less than whole-hearted. Throughout his account of animal generation, Gassendi continues to ascribe some agency in the creation of seeds to Nature itself: "Nature, having gradually become accustomed [to propagation], learned to procure the propagation of animals similar in kind, so that from the perpetual motion and ordering of atoms [nature] acquired a certain necessity to continually operate in this way" (2.274b). This passage is part of Gassendi's exposition of Epicurus' view. But on the next page, it becomes clear that he is willing to endorse the Epicurean claim, so long as one major caveat is made (2.275a). Namely, there must be some cause of the world's order that is external to atoms:

> These motions were first given to atoms or molecules and seeds of things by God, creator of the world. In this way we can grasp that the series and succession

of things, and especially the propagation of animals, proceeds as God himself instituted, so that the concatenation and dependence of motions on motions and things on things proceeds necessarily. In consequence of this, it can be said that nature acts as it does and in such a way that it could not act otherwise, given the course it began on and which it continues on. (2.275a)

The effect of this caveat is to eliminate the element of chance and thus to make Lucretian evolution far more acceptable. Gassendi insists that God created atoms with a fixed quantity of *vis motrix* and that the motive power of atoms fully determines the molecules and seeds that will come into existence some time after the creation.

At first reading, Gassendi's two competing suggestions, that God created an apparently chaotic mass of unstructured atoms and that God created the various species of animals, plants, and minerals found in the world today, seem wildly different. However, by transforming Lucretius's chance arrangement of atoms into one foreordained by God through his disposition of motive power, Gassendi brings the two quite close together. In Gassendi's evolutionist story, even if the initial creation appears chaotic, the chaos is merely apparent.

It is, I think, relatively clear that Gassendi prefers his version of Lucretian evolution, on which the divine creation of seeds is indirect. However, he is somewhat worried about the compatibility of such an account with literalist readings of the Bible. Gassendi's general epistemological pessimism is helpful to him here. Just as in the case of Copernicanism, we are presented with two possible explanations of the phenomena, both of which accord with reason and neither of which can be demonstrated. If providing a final answer is left to faith, this does not show that there is any conflict between the deliverances of faith and reason: It merely shows the weakness of the human mind, a theme Gassendi insists on throughout his career.

The Probability and Significance of Seminal Explanations

It may be helpful to compare Gassendi's account of generation with a more straightforwardly reductionist account like Descartes's. Stephen Gaukroger describes the Cartesian account of generation as an attempt to provide a new alternative to traditional views of generation and traditional physiology more generally. Some of Descartes's predecessors explained plant and animal physiology in terms of qualitatively different kinds of matter like the four elements, and this Galenist strategy was typically favored in medical faculties more than in philosophy faculties. A more straightforwardly Aristotelian strategy, favored mostly by philosophers, appealed to immaterial, formal principles like the vegetative and sensitive souls. A third strategy was teleological, and both medical authors and those from philosophy faculties used it.[23] Indeed, important works like Jean Fernel's *Physiologia* used all three. In contrast, Descartes's account of animal physiology is intended to avoid appeal to immaterial, formal principles, intrinsic teleology, or a fundamentally differentiated matter theory.[24]

Gassendi's account has anti-Galenist and anti-Aristotelian features, as his insistence that all structure is present from the moment of conception on shows.. However, he appears to use elements from all three strategies. Consider appeals to different kinds of matter. At bottom, atoms are homogeneous and vary only in size, shape, and motive power – but talk of the *flos materiae*, the most subtle and active sort of atoms, brings in different kinds of matter, as does his practical reliance on corpuscles rather than atoms. Gassendi does not use genuinely immaterial principles like forms, but he does use principles different in kind from the atoms that can be used to explain all nonliving things. The powers Gassendi describes as the *scientia* and *industria* of seeds, for instance, are not reducible to atoms and their motion. Rather, the powers of seeds are either superadded to them at the moment of creation – as the direct divine creation account suggests – or are emergent properties of atomic composites. This in effect introduces an essentially nonatomic power into Gassendi's ontology.

Finally, the *scientia* of seeds may be thought of as playing the role of a final cause in generation. Margaret Osler takes this as a prime example of the reintroduction of finality into physics.[25] I agree that Gassendi's account has the effect of appealing to natural teleology. He has no good explanation of how seminal power as an efficient cause can turn the epitome of the offspring into a developed organism, but it is easy to see how seminal power could be the final cause of this. However, I do not think that Gassendi intends his account of generation to appeal to genuine final causes aside from God's intentions. Were he to accept a form of primitive teleology, he would not be licensed in claiming that we do not understand how the seminal power operates. For if the power were primitive, there would be nothing to explain, and hence nothing

23 Gaukroger, *Descartes' System of Natural Philosophy*, 181–3.

24 However, Descartes's appeal to the three kinds of matter gives him the resources of a differentiated matter theory for all practical purposes.

25 Osler, "Whose Ends?" 159 ff.

lacking in our understanding once the power was noticed. But Gassendi *does* remind us of the mysterious nature of the seminal power:

> The *vis seminalis* and its industry, action, and mode of acting, the instruments that it uses, the material that it selects, its tractability and the order in which it is arranged, the separation and cohesion of the various parts among themselves... and hundreds of things of the sort elude all human perspicacity and sagacity. (2.172a)

And again,

> [w]hen these great endeavors of nature – endeavors at the elaboration of which we were neither spectators nor accomplices – are exposed to our view, we become completely stupid and like ignorant forest-dwellers we can only be struck dumb, not divining or determining by what artifice they have been completed. For indeed every one of them is a little machine in which almost innumerable little machines, each with their own little motions, are incomprehensibly enclosed. (2.267a)

Instead of fully comprehending how generation occurs, "what remains for us... is to sing a hymn to that divine and incomparable architect who has created and constituted within the seeds of things these, as it were, craftsmen, endowed with so much providence, industry and skill" (Ibid.). We are in the same situation as a man "born in the woods and ignorant of all human artifice." who, shown a tiny clock, "would simply marvel at the subtlety and elegance of its structure... and would never guess how the little machine could be made so perfect" (Ibid.).

The ignorant woodsman fails to understand something when he simply marvels at the clock. Namely, he fails to understand the mechanism by which it was produced. Similarly, when we marvel at the process of generation and end up praising God, we fail to understand the mechanism by which his works are produced. Gassendi's account of generation contains a gap because it does not spell out how the power that turns seeds into offspring operates. A complete account, however, would not need to advert to anything but atomic motions and the powers of molecules. It is a consequence of the limitations of the human cognitive capacities that we cannot give a complete, nonteleological account of generation.

These limitations have the fortunate consequence of drawing our attention away from the natural world to God. Whenever we realize that an efficient causal account of the phenomena at hand is beyond our grasp, and consider instead God's intentions in bringing those phenomena about, our knowledge and admiration of God is increased. Gassendi's skepticism about whether we can ever arrive at full knowledge of nature thus buttresses the argument from design and turns natural philosophy in the direction of God.

Notice that in turning our attention from nature to its author, we move from one order of causes to another. There is room for both an efficient causal account in terms of secondary causes and an account in terms of God's intentions. As we saw in Chapter 6, Gassendi has a number of reasons for denying that natural events are caused *solely* by God and not by creatures as well. The fact that Gassendi's natural philosophical explanations often end up pointing at divine intentions does not by any means show that he thinks God intervenes directly in nature as an efficient cause.

We shall see a very similar pattern in Gassendi's account of sensation as a power of the corporeal soul. He is confident that this account, on which sensation is carried out by a particularly fine and subtle body interspersed throughout the coarser matter of the brain, is basically correct. However, he is skeptical about whether a full account of how that subtle body produces sensory states could ever be given. Here, as in the case of generation, we are supposed to be able to understand *enough* about how the process works to know that Gassendi's account is better than its competitors, but not enough for us to be able to tell a satisfying story in terms of secondary causes alone. Given the limitations of the human cognitive capacities, we must bring divine action into the account. To see how this works, let us turn to the corporeal soul and its powers.

The Corporeal Soul

Gassendian human beings have two souls, a material *anima* composed of particularly fine atoms "by which we are nourished and by which we sense" and an incorporeal *animus* "by which we reason" (2.237b). The incorporeal soul is supposed to be joined to the body via the corporeal soul, which is "as it were the medium or nexus" joining the two (2.256b). Gassendi devotes very little attention to explaining how cognitive labor is distributed between the two souls, in part because he discusses the two in very different places. I examine the corporeal soul here, and defer its relations with the incorporeal soul until Chapter 10.

The scholastic textbooks Gassendi used when teaching explained that we can consider the soul either in respect of its nature or as it pertains to the faculties of the ensouled creature.[26] Discussions of the soul

[26] Eustachius, *Summa philosophiae quadripartita*, 165 (Section 3.1).

considered in these two ways are widely separated, just as they are in the *Syntagma.* Gassendi also discusses the traditional questions in the traditional order. Eustachius, for instance, asks how the soul should be defined; whether it has a particular seat or is joined with the whole body equally; whether there are different kinds of souls; whether there is more than one soul in any individual living body; and whether the soul is indivisible.[27] Looking at Gassendi's answers to these traditional questions is a useful introduction to his account of sensation.

The soul, for Gassendi, is the principle of living and sensing. Because it is not a form, it cannot be described as informing the body as a whole; rather, Gassendi says, the soul is "diffused" throughout the entire body. Nevertheless, different powers of the soul are localized. The sensitive power has its seat in the brain and nerves, and the heating power of the soul – which corresponds to the principle of life shared by plants and animals – is seated in the heart, at least in the case of animals with hearts. There is only one kind of corporeal soul, but it exists in more and less perfect versions. For instance, cow souls are more perfect than fish souls because cows but not fish have a well-developed internal principle of heat, and fish souls are more perfect than plant souls because they endow the body with locomotive and sensory powers. The souls of plants and animals differ only in degree of perfection, however, and not in kind.

It follows from the claim that there is only one kind of corporeal soul that each individual body has need of only one corporeal soul: Animals, for instance, do not have two really distinct souls but rather one soul with both vegetative and sensitive powers. Here Gassendi is in effect splitting the difference between an Aristotelian ontology on which organisms possess one soul with different aspects and a Galenist ontology on which organisms possess distinct souls seated in distinct parts of the body. His conception of the soul as a composite body allows him to maintain the uniqueness of the soul while localizing different functions in different parts of the body.

Gassendi's account of the *anima* begins with a familiar skeptical caveat: "there is no hope," he remarks, "that we might observe the innermost nature of the soul" (2.250a). However, taking the manifest life of an animal as a sign, we can form opinions about its inner cause, and Gassendi infers that "the *anima* is something extremely subtle, which, although it cannot be perceived by vision, nevertheless can be perceived by understanding and deduced by reasoning from the heating, nutritive, sensing,

[27] Ibid., 165–79 (Sections 3.1.1–10).

motive, and other functions that could not exist without a principle by which they are elicited" (2.250a). The *anima* is thus a theoretical entity, just as atoms and the void are. Moreover – at least to the extent that it is defined simply as the principle of heating, nutrition, and the other functions traditionally ascribed to the sensitive and vegetative souls – we know its existence with the same certainty as we know the existence of atoms and the void. However, Gassendi does not give us any real reason to think that it is one and the same entity that (for instance) heats and senses: This assumption is built into his definition of the soul.

Gassendi follows his typical pattern of argument in developing an account of the soul. He enumerates a number of contemporary and ancient options, rebuts or amends them all, and proposes his own account as the only remaining alternative. The two main points of his rebuttal are the following. The *anima* cannot, he says, be a form of the sort the Peripatetics describe because he has previously shown that there are no such forms. Nor can it be a mere quality, *habitus* or disposition of the parts of matter, as Aristotle and Plato are said to have thought. For a disposition or *habitus* is a mere relation, but the *anima*, the principle of various sorts of animal actions, must be something active (1.250b).[28] Thus, the *anima* must be a body. Recognizing that this opinion is unusual, since the great majority of philosophers have held the *anima* to be incorporeal, Gassendi offers a diagnosis of the common but mistaken view: Previous philosophers were misled by the great fineness and subtlety of the *anima*, which, in comparison with the coarse mass of the rest of the body, makes it appear as if incorporeal (2.250a–1b). Thus, the invisibility and intangibility of the *anima* are products of its corporeal subtlety rather than its incorporeality.

The *anima* is composed from the most subtle, mobile, and active corpuscles, which fire and heat are also composed from (2.250b). Indeed, the soul itself is "a certain little flame or most subtle sort of fire," a view that – bracketing out disagreements about whether fire, heat, and light are corporeal – Gassendi attributes to many of the ancients, including Plato and Aristotle themselves. He advances five distinct reasons by which "it can be established that the *anima* is a certain sort of fire" (2.251b). First, the fact that living, sensitive bodies are warm and dead ones cold suggests that the *anima*, the principle of life, heats coarse bodies, and fire is the chief cause of heating. Gassendi's initial limitation of his discussion to the case of breathing, sanguineous animals – dogs, horses, chickens,

[28] This, like the first rebuttal, relies on the criticisms developed in the *Exercitationes*.

and eagles – thus becomes crucial (2.251b). However, someone like Fernel explicitly argues that even though vital heat is less evident in plants and in animals like snakes, it can still be shown to be there.[29] Second, we can infer that the *anima* is a sort of fire "from its constant need and consumption of nourishment" (2.251b). Third, respiration also suggests that the *anima* is a sort of fire because fire requires air as well as fuel (2.252a). It is generally agreed that the heart is the cause of the continual motion of the blood, and Gassendi's suggestion is that it does so by heating the heavy and inflammable material of the blood. Fourth, the *anima*'s efficacy at moving the much heavier body suggests that it is a sort of fire: Fire, alone among things we observe, has the power to move things much heavier than itself (2.252a). Finally, Gassendi notes the continual motion and agitation of images within the fantasy, not only during waking life but also when we are asleep and dreaming. This sort of continuous motion is proper to fire, or more precisely flame (2.252b). None of these reasons are individually supposed to be conclusive; however, taken together with Gassendi's argument that the *anima* must be a body of some sort, they are supposed to give quite strong support to the view that "the soul is nothing other than a flame kindled within the body of an animal, which is its principle of vegetation, sense, and all other vital actions" (2.252b).

Gassendi bases his account of the sensitive powers on this view of the *anima* as a sort of fire. His explanation of the role of the *anima* in sensation begins by asking how, given that everything is constituted from individually insensate atoms, animals and their souls are capable of sensing (2.343a). This is not a question for the atomist alone; a more general version of the problem arises, as Gassendi notes, for any of the philosophers "[w]ho compose everything from a certain ungenerable, incorruptible, and quality-less matter, and likewise for those who compose everything from the common, chemical, or other elements" (2.343a–b). A philosopher who held that the soul arises from earth, air, fire, and water, or from salt, sulfur, and mercury would face the same question: How do sensate souls arise from insensate elements? For no one, so far as I know, thought that *all* atoms or elements possessed sensory faculties.

Gassendi's answer is that sensation arises when the soul is kindled in the body as fire is kindled in a log (2.345a). The comparison is intended quite literally: Gassendi writes that "food such as bread or herbs is no more distant from living and sensing flesh than a log is from a light-giving and burning flame." There is supposed to be no greater difficulty

[29] Fernel, *Physiologia*, 259 (Section 4.1).

in explaining how sensation can arise from matter devoid of qualities than in explaining how light and heat can arise from matter devoid of qualities. Because we have already accepted that light and heat *do* arise from matter devoid of qualities, we have no reason to deny that sensation can so arise as well.

The comparison continues as follows:

> Just as . . . particles can be disentangled from a log, which particles will have a new power of lighting and heating once they move and arrange and dispose themselves in a new way – so spirituous particles can be obtained from dissociated food, which particles will possess an *energeia* of sensing once they are divided in a certain manner and disposed in a new way. (2.345a)

Gassendi considers the objection that his account has failed to explain anything because, after all, "corpuscles of heat or flame considered as individuals . . . do not sense" (2.346b). In this connection, he mourns our lack of knowledge of even manifest things and notes "[b]y how much more we are deceived, by how much more it eludes all human understand to grasp the texture and temperament that can be judged to be the principle of sensing for the soul – or which parts or organs the soul . . . uses to sense" (2.346b–7a). He goes on to explain how much remains unknown even in the chemical processes involved when a log is burning. (He has in mind chiefly the various changes in the action of the salts contained in the log.) This comparison is supposed to show that there is no special difficulty in understanding how sensibility arises from insensible things. We know *that* sensation arises from a corporeal soul and the fact that we cannot explain *why* it happens should no more make us give up that claim than our inability to understand certain chemical processes should make us give up chemistry.

Gassendi goes on to list everything that remains unknown, namely the particular size, shape, motion, position and arrangement of the atoms that constitute sensate bodies. This list includes pretty much everything there is to know. But the provision of such a list implies that sensory power *does* emerge from arrangements of atoms with particular size, shape, motion, and position.

Everything we have seen so far is in keeping with this implication. However, near the end of his discussion of generation, Gassendi adds a caveat that ties together his accounts of sensation and generation. Retreating from the Epicurean line of argument dramatically, Gassendi explains that sensate things actually arise from *semina* that, although not themselves sensate, nevertheless "are or contain principles of sense" (2.347b).

This disolves the problem of explaining how sensate bodies can arise from insensate parts and replaces it with the much easier problem of explaining how a tiny sensory capacity can grow into a larger one. The claim that there are superadded or emergent powers is nowhere near so well integrated in Gassendi's account of sensation as it is in his account of generation, but it is nonetheless there.

It would be natural to ask, at this point, where the animate *semina* gain their powers from, but we have already seen that Gassendi gives two possible answers to this question. Either God created animate seeds that have the power to reproduce themselves (direct divine creation) or he created atoms endowed with such quantity and direction of *vis motrix* that they would eventually come together to form the appropriate kinds of seeds (Lucretian evolution or indirect divine creation). He leaves the choice between the two up to his readers, although, as in the case of Copernicanism, it is clear which choice he thinks is philosophically preferable.

9

The Metaphysics of Body

Gassendi's metaphysics is intended to preserve traditional, common-sense intuitions about substance within a materialist ontology of body. He articulates an anti hylemorphist account of substance and a reductionist account of the qualities of inanimate bodies that rivals, and in many ways resembles, the one Descartes provides. Unlike Descartes, however, Gassendi attempts to preserve the substantiality and individuality of the plants and animals we interact with. Thus, he needs to answer the sort of questions about unity and identity over time that were traditionally answered by appeal to form.

He does so in the course of providing a natural philosophy rather than in anything devoted to metaphysics so called. Although most of Gassendi's contemporaries accorded a prominent place to a discipline they called metaphysics, the *Syntagma* moves from logic to physics to ethics, with no book on metaphysics at all. This is the standard Stoic and Epicurean ordering, so it is unsurprising that Gassendi adopted it.

Rejecting metaphysics as an independent branch of philosophy does not imply anything like a rejection of the sort of ontological questions typically taken to be central to metaphysics today. Nor does it involve a rejection of the sort of natural theological questions that Gassendi's contemporaries often understood as central to metaphysics. For instance, he treats topics such as the nature of providence, the various proofs of God's existence, and the immortality of the soul as part of his physics, thus integrating them into what he considers to be first philosophy (1.133b–4a).[1]

In what sense of the term, then, does Gassendi reject metaphysics? He uses the term very rarely; in fact, I know only two interesting cases. In the preface to the *Exercitationes*, he describes his plan for the sixth book of that work as follows:

> Book VI is directed against metaphysics. This book, in which the greatest part of the maxims offered for metaphysics are rejected, powerfully attacks its commonly known principles and its famous properties of being, unity, truth and goodness. Then it attributes whatever knowledge is possessed concerning intelligences and the thrice great God to orthodox faith. For it shows how truly vain are those arguments by means of which [philosophers] are accustomed to philosophize about separated substances from the natural light. (3.102, no columns)

"Metaphysics" in this sense concerns God, angels, and the human soul after death, as well as the abstract properties of being, unity, truth, and goodness. In the *Exercitationes*, Gassendi planned to argue that we can know God and the angels only through faith and to set aside the remaining metaphysical issues as irrelevant. The second part of this strategy persists in Gassendi's mature work, but by the time of the *Animadversiones* he abandons the claim that we cannot come to know God through reason. In his final work, he adopts a somewhat more complex – and far more orthodox – account of the relation between faith and the natural light, thereby integrating metaphysics into physics.

Gassendi suggests that Aristotle himself never intended to establish an independent discipline of metaphysics. Aristotle's followers, he says, called topics concerning separated substances and general reasons metaphysical simply because they were treated in the book that followed the *Physics* (1.27a). Thus, in typical humanist fashion, Gassendi discredits the various Aristotelians of his day by challenging their claim to provide the proper interpretation of Aristotle and suggests that in fact his philosophy is more in sympathy with Aristotle. This claim is made in the Proem to the *Syntagma*, where Gassendi outlines the proper way of dividing philosophy up into disciplines. Aristotle, he says, divided contemplative philosophy into mathematics, physics, and theology, where theology "contemplates divine things and things separated from body, and also the general reasons of all things or of being in general" (1.27a). But the

[1] More precisely, the first, programmatic part of physics – which concerns topics like space, time, atoms, and motion – is first philosophy, the science that deals with being and nature in general.

Stoics and Epicureans combined theology with physics, "since the divine nature makes itself known from the fabrication and governance of the universe" (1.27a). Now for Gassendi this cannot be the end of the story because the divine nature also makes itself known through revelation. But the Stoic and Epicurean move is echoed in his endorsement of the argument from design as the best, and indeed only, rational demonstration of God's existence and the best rational incentive to worship him.

Thus, by rejecting the discipline of metaphysics, Gassendi is not rejecting the type of questions we now count as metaphysical and that, as a result of the work of figures such as Descartes and Leibniz, had come to count as metaphysical by the end of the seventeenth century. Gassendi is deeply engaged with ontological questions that count as metaphysical in this later sense of the term. Indeed, the *Disquisitio* and the *Syntagma* make a number of important ontological claims and assumptions. It is worth reconstructing these ontological claims in a systematic fashion to help us see the continuity between Gassendi and his successors. I focus on three topics where Gassendi's and his successors' reinterpretation of Aristotelian ontology is clearest: the notion of substance, in particular the unity and diachronic identity of substances; the related notion of essence; and the ontology of qualities, in particular the transformation of the Aristotelian distinction between manifest and occult qualities into the distinction between what will later be called primary and secondary qualities.[2] I begin with substance.

The Role of Substance

Substances are the fundamental entities for philosophers from Aristotle onward. Thus, providing an account of substance is a crucial task for anyone who, like Gassendi, rejects the Aristotelian conception of substance as informed matter. Gassendi's intention is not to provide a deliberatively revisionary account of substance but rather to make clear how something fulfilling the same desiderata as Aristotelian accounts can be provided within an atomist framework. When he insists that there is nothing to substance over and above matter, there must be enough similarity in the two conceptions of substance that he is not simply changing the subject.

[2] Gassendi does not use the terms primary quality and secondary quality in this sense. Rather, he adopts an Aristotelian usage according to which the primary qualities are the elemental qualities warmth, coldness, wetness, and dryness, and the secondary qualities are such qualities as rarity and density. For a full account of the Aristotelian usage, see Goclenius, *Lexicon philosophicum*, 912 ff.

His main conceptual innovation concerning substance is the postulation of space and time as genuine beings that are capable of existing independently without being substances, thus teasing apart the two standard criteria of substancehood: independence and being the subject of properties.

For Gassendi, as for Aristotle, plants, animals, and human beings are the paradigmatic substances. However, they are not the *only* ones: God, the incorporeal human soul, inanimate bodies and – occasionally – even individual atoms (1.372a) all count as substances. It is a difficult question whether Gassendi's considered view is that atoms are substances. If they were not substances, then a substance such as a stone would be composed of parts that can exist independently but are not themselves substances, and this is an unappetizing conclusion. However, because atoms do not have essences, they do not fit entirely comfortably into the category of substance.

One requirement for an acceptable theory of substance is that it must explain how substances are the fundamental ontological entities and what the unity of a substance consists in, that is, what makes a certain group of properties into one thing. A hylemorphic account of substance would say that chunks of matter are unified in virtue of being informed by a single form, but what is the parallel atomist answer? What makes it true that a collection of atoms is one substance? The two answers that suggest themselves immediately are both unsatisfactory. One obvious suggestion is that substances are individuated by their souls. This would explain how paradigmatic substances like plants and animals are unified. However, it does not explain how metals and minerals, which Gassendi also counts as substances, are unified. More seriously, it just seems to push the problem back a step. A corporeal soul is just a congeries of atoms, so what unifies *it*? A second obvious suggestion is that substances are individuated by their textures – but this raises a similar problem. What grounds our division of the world into different textures? Why not construe the whole world as having *one* texture and hence being one substance, as Spinoza does?

Similar problems arise concerning identity over time. Gassendi recognizes that animals gain atoms in nutrition and lose atoms in elimination and generation, so sameness of atoms cannot explain identity through change. Because textures and corporeal souls change over time, we cannot appeal to them to explain the diachronic identity of substances without already having a criterion for textural identity. Indeed, on Gassendi's account of the relation between textures and the qualities they give rise to, qualitative change *requires* textural change.

Gassendi never explicitly answers the questions of individuation that immediately arise for an atomist theory of substance. His interest in substance is rather more limited, as can be seen by examining the three different conceptions of substance put forward in the *Syntagma*:

(1) Substance is that which qualities flow from, the "stuff" of which qualities consist. This conception, which implies a denial of real accidents, leads to the equation of a substance with its nature or essence (1.183b, 3.312b) and is involved in Gassendi's frequent contrasts between the "hidden," "inner" substance that we do not know and the qualities or appearances we sense (3.310a, 3.369a). One chief source of Gassendi's interest in the notion of substance is this contrast between substance or essence and appearance.

(2) Substance is that which exists per se (1.179b, 1.182a). This definition amounts to an endorsement of the claim that substances are the basic constituents of the created world; substances depend on nothing other than divine conservation, but qualities depend on substances. The equation of substantiality with independence and accidentalness with dependence precludes space and time being accidents, and another chief source of Gassendi's interest in the notion of substance is to show that space and time are neither substances nor accidents thereof.

(3) Substance is that which acts and is acted upon (1.184b). This is the conception of substance that Gassendi's account of space and time relies on the most, and it precludes space being a substance. But although Gassendi says that substantiality requires both action and passion in the course of discussing space, it is clear from what he says elsewhere that he should instead say that substance acts *or* is acted upon. Because God is a substance who cannot be acted upon, it must be possible for an impassive substance to exist.[3] And because there may be atoms with no *vis motrix*, it must be possible for an inactive substance to exist. Moreover, when Gassendi attacks the Cartesian conception of body as inert, extended substance, he does not attack the *coherence* of that notion but rather its ability to explain motion and change.

[3] Although many philosophers have held that substance is equivocal between God and created substances, Gassendi cannot accept this view on pain of making it impossible for us to conceive of God as a substance. See Chapter 10.

These three conceptions of substance are neutral as to what sorts of substances actually exist. They would allow for substances composed of matter and form, immaterial substances, and entirely material substances. As such, Gassendi intends them to apply to God, atoms, and the human soul as well as to bodies. However, he devotes rather more attention to the ontology of corporeal substance, offering a corpuscularian account of (1) on which qualities are reducible to the texture of corpuscles composing a body:

(4) Substance is matter, for there is nothing in bodies but matter.[4]
(5) Substance is nature. Gassendi very often speaks of the "hidden substance" or "inner nature" of things as unknown. At points he makes the stronger claim that substances or natures are *unknowable.*

In the case of inanimate composite bodies, the substance or nature of a body is its texture of atoms, and that texture gives rise to all the qualities of the body. The distinction between substance or essence and appearance is the distinction between the corpuscularian texture constituting a thing and the different ways that texture affects different perceivers in different circumstances. Thus, the unknowability of substance is in the first place a matter of its being hidden from perception. However, Gassendi also has a further concern about the unknowability of natures.

Essences or Natures

We know something about how natures or essences are conceived by knowing that they are commonly identified with the substances they belong to. More about the role essences play becomes apparent when we see what it is that Gassendi thinks we would know if we knew the essences of things. In the *Exercitationes,* Gassendi – questioning the Aristotelian claim to knowledge of natures – makes clear that if we knew the natures of substances such as horses and apples, we could answer the following questions:

How thick or rarefied is [the horse's soul]? How does it reach and then penetrate so many different parts? How does it bring sensation to them? Or imagination? And for what reason does this soul make necessary a certain type of head, a certain

[4] For instance, Gassendi glosses the claim that generation produces no new substances, only new qualities or modes, as the claim that generation creates no new atoms (1.472a). More commonly, he simply equates substance with matter (1.184b, 1.372a) or with groups of atoms (1.372a, 1.373b).

type of feet? . . . What is the *energeia* that in the first place drives roots down from such a little granule and leads forth a trunk and so many different branches? What then brings nourishment to the furthest branches with so much providence, and gives just as much as is necessary to each of the parts? (3.185b)

This is from the *Exercitationes*, an early text. But very similar language appears in the *Disquisitio*, where Gassendi is challenging Descartes's claim to know the essence of the mind:

When you say that you are simply a thing that thinks you mention an operation which everyone was already aware of – but you say nothing about the substance carrying out this operation: what sort of substance it is, what it consists in, how it organizes itself in order to carry out its various different functions in different ways, and other issues of this sort, which we have not known about up till now. (3.306b)

He reiterates the question, adding more detail about what he wants to know:

What sort of thing [the underlying *principium* of thought] is, how it exists, how it holds together, how it acts, whether it has certain faculties and functions, whether or not it has parts, and if it has any, what kind they are: if it does not have any and is indivisible, how it arranges itself in so many different forms; how it performs so many functions; by what means it deals with the body; by what means it goes beyond it; how it lives without it; how it is affected by it. (3.306b)

The fact that Gassendi places such similar demands on the otherwise very different notions of essences that Descartes and scholastic Aristotelians use provides evidence that he is not simply pointing out internal requirements of those theories but rather explaining what genuine knowledge of essences would provide.

We do not, then, come to know essences using Aristotelian logic and natural philosophy, nor by inspecting our allegedly clear and distinct ideas. But how *could* we come to know essences? Here Gassendi's answer is less obvious. In fact, he tends to suggest on the one hand that knowledge of essences could be attained through empirical investigation, and on the other that we should not expect to ever attain such knowledge.

The first suggestion is made very clearly in the *Fifth Objections's* famous demand for a "chemical investigation of the mind":

Given that you are looking for knowledge of yourself that is superior to common knowledge . . . it is certainly not enough for you to announce that you are a thing that thinks and doubts and understands, etc. You should carefully scrutinize yourself and conduct something like a chemical investigation [*quasi chymico labore*] of yourself, if you are to succeed in uncovering and explaining to us your internal substance. (3.311a)

Uncovering the nature or substance underlying the manifest properties of the mind requires some sort of empirical investigation. Similarly, physical explanations involve discerning the particular corpuscularian structures of the things under investigation and seeing how these structures produce the manifest features of the thing. These underlying structures are natures or essences (2.463a).[5]

Now, even if we dismiss Gassendi's suggestion that natures are *in principle* unknowable, he clearly thinks that natures are unknown *at this time.* We have not yet succeeded in giving explanations that reach all the way down to structures of atoms and the void; rather, the "secret or hidden essences" lying behind the appearances remain obscure to us (3.352b). However, we have met with some success in explaining manifest behavior in terms of the behavior of insensible parts. An example of such success may be helpful:

> We used to marvel when people pointed out that a small piece of rotten cheese, strewn on a garment and brought near the skin, caused so much stinging to the senses. The Microscope has demonstrated the reason, namely, that all those grains [of cheese] are little animals who, among other things, push their little beaks into the skin and bore through and damage it, just as they might bore through and damage the surface of the earth, in order to seek their food. (2.463a–b)

Here, we have explained an operation of a composite body in terms of the behavior of smaller parts making up that thing. This sort of explanation is the first step on the way to a complete natural philosophical explanation. We should then investigate the "little animals" themselves, and so on, until we arrive at an explanation in terms of atoms.

The suggestion that human beings are precluded from knowing essences is one that Mersenne makes in terms very similar to Gassendi's, and it is accorded central importance in Popkin's characterization of Mersenne and Gassendi as mitigated skeptics.[6] It is tempting to try to reconcile the suggestion that essences are in principle unknowable with

[5] However, Gassendi sometimes also uses the terms essence and nature to refer to the ideas such types produce in the mind. This usage is most common in cases where Gassendi addresses the view that he ascribes to both Descartes and the Aristotelians, the view that essences are immaterial and possess some mind-independent reality. Thus, Gassendi uses essence in the sense of idea to emphasize that *Cartesian* or *Aristotelian* essences cannot be anything more than ideas (3.377a).

[6] For the suggestion that humans cannot know essences, see Mersenne, *La vérité des sciences*, 9.

the suggestion that essences are unknown just in virtue of the deficiencies of current natural philosophy. In the first claim, one might suspect, essences are essences as the Aristotelians or Descartes understood them; this is why the first claim is so much more prominent in the *Exercitationes* and *Disquisitio* than the *Syntagma*. In the second claim, essences are corpuscularian structures. The inconsistency is merely terminological.

Now there clearly is some terminological inconsistency, but I suspect there is more going on as well. For on Gassendi's view, essences understood as corpuscularian structures are not the sort of thing that could be grasped by sense perception, even if they could be grasped by some other means. Consider the opposition between essence and appearance, that is, between hidden substance and manifest properties:

> Because an attribute or property is one thing and the substance or nature of which it is [an attribute] or from which it emanates is another, to know an attribute or property or collection of properties is not thereby to know the substance or nature itself. What we can know is this or that property of this substance or nature when it lies open to observation and is perspicuous in experience, and we do not thereby penetrate into the inner substance or nature – just as when looking at bubbling spring water, we know that the water comes from a certain source but do not thereby direct a keen look into the interior and discover that subterranean source. (3.312b)[7]

When we look at the spring water rising from the ground, we know that there is an invisible, subterranean spring because water bubbling from the ground is a commemorative sign of such a spring. But we do not know exactly what the hidden spring is like; we have no clear idea of it but simply imagine it on analogy with the visible springs we have seen. Such analogical ideas can never be as certain or as informative as ideas acquired through sense, for what is known by the use of indicative signs does not transcend probability and our knowledge of the manifest is, in the best case, entirely certain.

The contrast between knowledge of appearance and knowledge of hidden substance, essence, or nature is also spelled out in terms of a contrast between knowledge of existence and knowledge of essence. This

[7] Compare the following passage, which follows Gassendi's suggestion that just as microscopes have allowed us to explain certain phenomena in terms of their constituents, we might be able to explain *all* phenomena if we could see their atomic parts: "[B]ecause we are destitute of vision of this sort (that is, microscopic vision), and because there is no great hope of ever obtaining such a splendid microscope . . . shouldn't we, I ask, content ourselves with those things our author equipped us to know, because [that knowledge] alone is necessary for us?" (2.463b).

approach is most prominent in the debate with Descartes. Arguing against Descartes's claim that we cannot cognize a thing without cognizing its essence, Gassendi writes that

[k]nowledge of the existence of a thing has no necessary connection with knowledge of its essence or inner nature, for otherwise we would know the nature, essence and inner depths of anything that was obvious to the senses or whose existence we knew of in any way at all. (3.290a–b)

Knowledge that a thing exists is cheap, on Gassendi's view: We just need to look at the thing. When someone senses a collection of accidents, they thereby naturally "conceive that there is something that is the subject of the accidents" (3.290b), and thus come to know that a certain thing exists and has accidents without acquiring a clear idea of it. Knowledge of essences or underlying substances is more difficult to acquire because it "requires a certain complete internal examination" so that "essence does not become known except by bringing to light every inner depth" (3.311b, 3.312a). The mere fact that essences are not perceived is supposed to show that knowledge of existence does not amount to knowledge of essence. The only way we could have knowledge of essences that is as certain as our knowledge of existence is if, *per impossibile*, we could literally perceive essences (2.463b). Thus, knowledge of sensible qualities will always have a certainty that knowledge of essences lacks.

Qualities

Let us now turn now to qualities, discussions of which occupy a rather large portion of the *Syntagma*'s *Physics*. Gassendi begins by explaining why we need to examine the nature of qualities in such detail:

Qualities are . . . that on which whatever reasoning is possible in physics depends. For all reasoning has its origin from the senses or from that which the senses perceive. But nothing beyond qualities is perceived by the senses because a quality is what appears to vision, touch, and the other senses. (1.372a)

Sense perceives only individual qualities, not the underlying substance from which they flow or the type they instantiate, and the intellect can universalize our sensory grasp of qualities only on the grounds of similarity. Gassendi's nominalism thus serves as one constraint on his account of qualities.

A second constraint is provided by the doctrine of the truth of the appearances, for that doctrine demands a way of understanding qualities on which the senses cannot possibly misrepresent them. Earlier, we saw how the *Exercitationes* used Sextus Empiricus' arguments against knowledge of qualities. Noting that honey tastes differently to a sick man and a healthy man, Sextus attempts to produce in his readers a state of mind in which they will suspend judgment about the true nature of honey. The *Exercitationes* concludes from such cases of perceptual relativity that we cannot know the inner nature of things: I can know, for instance, that honey tastes sweet to me but not whether the honey is *in itself* sweet or not.[8] Responding to his own early skepticism, in the *Syntagma* Gassendi provides an atomist explanation of how honey appears differently to different perceivers despite its unchanging nature. Thus, the perceived variability of the taste of honey raises no skeptical problems for the variability is in the perceivers (or in the chain of transmission from object to perceiver) and not in the honey.

It is important to see that Gassendi does not merely offer the sort of restricted dissolution of perceptual relativity that Locke offers in the case of the secondary qualities. He addresses the variability of perception of primary qualities as well. The alternately square and round appearances of the same tower provide no reason to doubt that the tower is in itself square, once we understand how the relevant *simulacra* are affected by their transmission through the air.

A third and final constraint is provided by Gassendi's atomism. I have argued that we should understand Gassendi as being committed to emergent or superadded powers in the case of substances endowed with souls and produced by generation. However, the qualities of nonliving bodies must ultimately be reduced to the qualities of atoms, either singly or as a group. In most cases, Gassendi will account for qualities in terms of textures of atoms, so that both the qualities of the individual atoms and their arrangement in space will be relevant.

With these constraints in place, let us turn to Gassendi's account of qualities. Gassendi begins by offering us a choice of definitions:

> Quality in general can be defined as the way of situating itself [*modus sese habendi*] of a substance; or as the state and condition in which material principles, having been mixed together among themselves, are situated; but it can also be defined paronymically, following Aristotle, as all that by which a thing is said to be such [*quales*]. (1.372b)

[8] Sextus would not, of course, make the further step to the dogmatic claim that whether honey is sweet in itself is unknowable. In this respect, the skepticism of the *Exercitationes* is more Academic than Pyrrhonian.

The first of these definitions is Epicurean and the second Stoic; the third, as Gassendi notes, is Aristotelian.[9] Now although there is a set of things captured by all three definitions, the definitions are not equivalent. The first two definitions are much broader than the Aristotelian one. They include things that would fall under categories other than quality in the Aristotelian schema, where what is predicable of substance also includes things under the categories of quantity, relation, place, time, situation, condition, action, and passion. Gassendi's account of space and time removes place and time from the scheme of categories (1.374b). He discounts the category of relation on the grounds that relations are "not in things themselves" but exist "only from an operation of the intellect" so that they are not properly *of bodies* at all (1.374a). The remaining categories can all be made species of quality because they all embody different ways of answering the question, "what sort of thing is this?" Thus, Gassendi insists, "whatever is observed in physical bodies (the rational soul in man excepted) is either a substance, which is one and the same thing as the body, the conglomerate of material or corporeal principles; or it is a quality, an accident, or a mode of the substance" (1.373b). Situation and condition, for instance, are qualities.

Although Gassendi talks about observing substances as well as their qualities in the preceding passage, it is clear that this is loose talk: We do not, strictly speaking, see substances but only their qualities. We may have mental awareness of substance, but sensory awareness of it is impossible. The claim that we do not, strictly speaking, perceive substances themselves was common, but it is interesting that Gassendi takes it to be a consequence of atomism:

> Atoms are the material principle that every composite or concrete body consists in. Thus, because atoms are the whole matter, substance or body that is in bodies, if we conceive or notice that something exists in these bodies, it is not a substance but only some mode of a substance, that is, a certain contexture, concretion, composition or consequence of matter or material principles. (1.372a–b)

Gassendi infers, from the premises that atoms are the substance or matter of bodies and that atoms are imperceptible, that substance is imperceptible. This seems like a non sequitur. After all, the most obvious reason that atoms are imperceptible is just their small size – and if this is the reason, then sufficiently large collections of atoms should still be perceptible.

[9] Compare Diogenes, *Lives of Eminent Philosophers*, 10.54–6 and Lucretius, *De rerum natura*, 2.333 ff.

However, presumably the point is supposed to be the familiar one that perception of an object is at best perception of it as qualified in some way so that it is not perception of the substance of the object itself but perception of the object as disposed in a certain way. We perceive color and the colored body, but "that which we denominate substance, being *in se*, we construct through induction, by which some subject that exists under the quality can be reasoned about" (1.372a).

In the passage just quoted, Gassendi writes that "everything we notice" in bodies is a "mode" rather than a substance (1.372a). He is rather careless with terminology, switching back and forth between the terms mode, accident, and quality for no apparent reason, but as the two passages just quoted make clear, everything in bodies that is not their substance is a mode of that substance. All qualities and accidents, properly understood, are modes.[10]

Traditionally, a mode is simply a way of being of a substance. On Gassendi's definition, a mode – at least, a mode of body – is a "contexture, concretion, composition, or consequence of matter or material principles" (1.372b), that is, a certain arrangement of atoms.[11] This by itself is a programmatic claim that tells us nothing about what the relevant contextures are, and the task of identifying the textures constituting particular modes has to be carried out on a case-by-case basis. The bulk of Gassendi's account of qualities in the *Syntagma* is devoted to carrying out that task.

Chapters are devoted to qualities such as rarity and density; magnitude, figure, color, and mobility; perspicuity and opacity; sharpness and smoothness; heat, cold, fluidity, firmness, wetness, and dryness; hardness, softness, flexibility, and ductility; and finally sympathy, antipathy, magnetism, electricity, and the other qualities called occult (1.372b, 1.375a). Gassendi attempts, with varying degrees of plausibility, to explain how they are all modes of variously textured composites of atoms. These explanations fall into different categories corresponding to the different types of qualities:

> [There are] qualities depending on [the atoms and the spaces between them] or adjunct to individual properties, of which sort are magnitude or quantity and

[10] This does not apply to the emergent or superadded powers relevant to generation because they are not, properly speaking, qualities.

[11] Gassendi never refers to modes of the immaterial mind. Indeed, for reasons that shall become clear in the next chapter, he never develops any of the ontological categories as they apply to the immaterial mind.

consequently subtlety and bluntness: and similarly figure, and what follows from it . . . and again motive power, the active faculty in general. . . . Or [there are qualities which are] adjunct to composite things, or things taken all at once: of this sort are first those that pertain to the various senses, as in respect of touch the qualities heat, cold, fluidity, firmness, humidity, dryness, and arising from them hardness, softness, flexibility, ductility, tractability and similar others. . . . Finally, [there are] qualities that pertain to interior faculties and less sensible operations, such as sympathy, antipathy, magnetism, electricism, and in a word all those said to be occult. (1.375a)

This is the main, tripartite division of qualities.[12] The first sort – size, shape, and their consequences, such as sharpness or bluntness – can be ascribed to atoms individually as well as in groups. This same group was called by Aristotelians the common sensibles and was soon to be called by Locke the primary qualities.

The question of how the size and shape of composite bodies relate to the size and shape of atoms has some interest. One might wonder, for instance, whether the terms size and shape are used univocally in the two cases. Because we derive our idea of the shape of an atom from our idea of macroscopic, apparent shape, there must be a reasonable degree of commonality between the two for us to even be able to entertain the notion of an atom. However, there may be differences as well. Most importantly, one might wonder whether the size and shape ascribed to composite bodies is a matter of appearance or not. This point is obscured by the fact that Gassendi uses the same word to describe macroscopic, apparent shape, and the shape of atoms. However, it seems clear that he would have to say that visual, tactile shape is an appearance and that composite bodies also have a real shape as well. This, indeed, is the chief point of difference between the first sort of qualities and the second and third: Only qualities like size and shape really apply to bodies in themselves as well as to bodies in respect of their causal powers.

A few points stand out in Gassendi's relatively brief discussion of this first sort of quality. Bodies need not have the same shape their component atoms do, but the mere fact that atoms have magnitude and are not mathematical points determines that all bodies have some shape (1.380a–b). Indeed, the main point of Gassendi's discussion of size and shape is not to explicate any complicated theory of the relation between atomic and

[12] Gassendi also sometimes hints that there is a *fourth* type as well, the sort exemplified by rarity and density. These are simply a matter of the spacing between atoms, and so, in a sense, qualities such as magnitude and figure derive from them. However, I am not sure what is supposed to differentiate the fourth sort of quality from the first.

observable size and shape but instead to press home the point that "extension is a mode of matter, or if you wish matter itself to the extent that it is not in a point but has parts beyond parts" (1.380b).

This raises a worry concerning transubstantiation that would have been obvious and urgent to Gassendi and his contemporaries. If extension is just a mode of matter, then how can the substance of Christ be really present in the Eucharist without his extension? And how can the extension of the bread remain after the bread itself disappears? To answer these questions, Gassendi argues that although it is a law of nature that bodies occupy places, having parts beyond parts, God need not obey this law in the Mystery of the Eucharist (1.381b).[13] For the laws of nature are "constituted by God's will," so his power is not limited by them (1.381a).[14] The possibility of transubstantiation thus relies on God setting aside the order of nature.

Gassendi's claim that extension is a mode of matter contrasts with both the scholastic construal of quantity as a distinct category and the Cartesian insistence that extension is the essence of matter. We have already discussed Gassendi's accusation that the Aristotelian scheme of categories amounts to reifying the various ways we conceive of substance. Against Descartes, Gassendi points out that it is solidity, not extension, that implies that two bodies cannot be in the same place at the same time. As we have seen, he takes the impossibility of co-location to be a criterion of body. He needs to deny that merely being extended implies the impossibility of co-location in order to allow bodies to be located in the void.

Let us move on to those qualities that exist only at the level of composite bodies. Gassendi divides these qualities into two sorts: those that affect "the various senses" – as heat affects touch, light affects vision, and so on – and those that pertain to "interior faculties" and "less sensible operations." Gassendi later defines this second group, the occult qualities, as those pertaining to "unknown faculties [*facultates incognitas*]" or "unknown causes [*causas incompertas*]" (1.449b). This might at first seem like an entirely natural switch from the insensibility of (presumably physical) interiors to the unknowability thereof. But in fact the "unknown

[13] I have not found in Gassendi an answer to the other common seventeenth-century worry about transubstantiation, namely, how the accidents of the bread remain after the substance of the bread has been replaced by the substance of Christ's body. Descartes, for instance, poses both questions in an undated letter to an unknown correspondent (*Oeuvres de Descartes*, 4.374–5).

[14] This is one of very few passages where Gassendi uses the term *lex naturae*.

causes" referred to must be the unknown causes that make occult qualities occult *on the scholastic view*. For on Gassendi's view, the causes of occult qualities are no more and no less well known than those of sensible qualities. The discussion of gravity in Chapter 7 goes some way toward showing this, but to make it clear we also need to look at an example of how Gassendi treats sensible qualities. Let us consider heat.

Heat is a somewhat complicated example because the term refers both to the object of a certain type of sensation and to a power to produce certain types of effects on other bodies.[15] However, all the examples turn out to be complicated in one way or another. Gassendi begins his discussion of heat by saying that heat – like the other three elementary qualities, cold, wetness, and dryness – "cannot be explained unless by the size, shape, and mobility of atoms" (1.394b). More specifically, he goes on to say that we feel the sensation of heat when heat enters into the pores of the body and penetrates and loosens its various parts, to which explanation he adds that when he says that heat "enters in, penetrates, loosens, etc., do not understand a certain nude and solitary quality but rather certain atoms insofar as they are furnished with such a size, such a shape, such a motion" (1.394b). Atoms possessed of such a size, shape, and motion are hence referred to as heat atoms.[16] In other words, atoms are heat atoms in virtue of being the sort of atom that, when they affect a human body, leads to the sensation of heat. Atoms themselves, properly speaking, have no heat: "Indeed these atoms do not have heat *ex se* or, what is the same

[15] Gassendi's chapter on heat focuses on how heat affects other concretions. This makes it somewhat mysterious why it is identified as one of the qualities "which pertain to the various senses." It seems that heat ought to count as both the second and the third sort. Gassendi first counts heat as sensible and then as occult, perhaps because he recognizes this, or because he does not see any difference in kind between the two sorts of effects just mentioned. Perhaps heat creates the sensation it does in virtue of "entering into and loosening connections" in the sense organ. This fits well with the apparent heterogeneity of the effects ascribed to heat, but it fails to explain how Gassendi can legitimately assume that all these different effects stem from the same underlying cause. At any rate, discussion of heat is complicated by its multiple roles in Gassendi's various explanations – something we might expect given the way heat is located in the various traditions he draws on, as, for instance, an elemental quality, the ultimate principle of life, and the source for the fundamental processes of alchemy (1.394a–401b).

[16] Given that Gassendi, like Sennert, sometimes calls corpuscles atoms, it is possible that he thinks of heat as a corpuscle rather than an atom. He sometimes speaks of heat-atoms as pyramidal, which would suggest that heat is a matter of noncomposite atoms; this is a neo-Platonist notion. On the other hand, heat and light are very closely allied for Gassendi, and light is clearly a matter of corpuscles rather than simple atoms. Moreover, it is hard to see how Gassendi can maintain that *all* qualities are modes if he really thought that heat pertained to individual atoms.

thing, they are not hot, but can nevertheless be judged and called atoms of heat, or calorific atoms, insofar as they create heat, that is, have this effect" (1.394b). One might wonder *why* or *how* such atoms create heat, but Gassendi's goal here is not to answer that question. To the extent that this is a question about why a particular atomic cause leads to a particular sensation, Gassendi is committed to its unanswerability. His concern is rather to make the point that a certain class of atoms is identified as hot in virtue of being the cause of a certain class of effects. Here he has in mind both the feeling of heat and the various burning, fermenting, and vivifying effects he takes to be produced by the same underlying cause. This fits in well with Gassendi's way of thinking about qualities as dispositions of atoms that cause the effects we experience, a point that will become clearer if we contrast the case of heat with its correlative, cold.

Gassendi ends his discussion of heat by asking "whether cold is a genuine and positive quality or a mere privation of heat" (1.401b) and answers that cold should be considered a positive quality just as heat is. We consider heat to be "a genuine, positive and active quality" because "if you set a glass bottle next to a coal the water contained in that bottle is so transmuted that it becomes hotter and bubbles." But "if you surround a glass bottle with snow or ice, the water is so transmuted that in a moderate amount of time it becomes colder and freezes." Thus, cold should be considered a positive quality, too (1.401b). For Gassendi adopts the general principle that a privation is incapable of action and concludes that cold must be a "genuine, positive and active quality" (1.401b).

The equation of being a genuine quality with having effects provides a useful clue to understanding the ontological status of qualities. For it suggests that for a body to have a quality simply is for the body to be the kind of thing that has a certain set of effects. Those aspects of texture in virtue of which the body produces a certain set of effects on the perceiver are its various sensible qualities.

This is, I take it, Gassendi's considered view. I admit that Gassendi's claim that heat atoms "are not hot" will be difficult to make sense of on this way of understanding the sensible qualities. If heat is whatever causes a certain sensation rather than the sensation itself or something that resembles it, then it cannot fail to be hot (that is, to cause the sensation). However, the claim that heat atoms are not hot will be difficult to understand literally on any account. It seems that Gassendi must simply mean that the atoms he calls heat atoms do not possess the scholastic real quality of heat. And this is all of a piece with his construal of heat, like the rest of the qualities, as a mode.

The definition of sensible qualities as atomic textures in virtue of which bodies produce effects carries over to the case of the occult qualities quite smoothly. However, Gassendi's account of the status of the occult qualities is rather clearer. Occult qualities are known by their effects and first picked out by means of those effects. We gain our concept of a magnet by observing that iron is pulled toward a particular group of bodies. We then imagine that there is something in these bodies that has that effect – the *vis magnetiva.* This occult quality, the magnetic power, is thus first known by its effects. Moreover, it is individuated by those effects: The occult quality is whatever it is in the body that gives it these effects. In the case of the sensible qualities, there is some semantic confusion about whether terms like heat apply primarily to the sensation or to the cause, and this makes Gassendi's account somewhat difficult to understand. There is no such confusion in the case of the occult qualities, where effects lack phenomenal character. There Gassendi, in keeping with ordinary language, is clear that the occult quality is the power in the body that produces a certain class of effects and not the effect itself.

In some passages, Gassendi contrasts occult qualities with sensible qualities. In others he has a trivision between occult qualities, sensible qualities, and "manifest" qualities, namely insensible qualities whose causes are known (1.449b). It is helpful to note this because even though the equation of sensible and manifest qualities can seem easy and natural from the perspective of the Aristotelian tradition, it is in fact quite mysterious why Gassendi would be willing to equate sensible qualities and manifest qualities, for one moral of Gassendi's atomist account of heat is that we previously did not understand what it is that we perceived by sense. Heat is only truly manifest if we understand its atomic causes. Gassendi conflates the unknownness and imperceptibility of occult qualities because of his semantic confusion over whether the quality-name applies to the cause or its effect. The *effect* can be considered occult whenever it pertains not to the senses but to another, nonsensate body. This yields a distinction between sensible and occult qualities on which Gassendi's manifest qualities will count as occult. But the *cause* can only be considered occult in virtue of being ill-understood. This yields a distinction between manifest and occult where the sensible qualities become manifest when they are properly understood.

On an Aristotelian picture, sensible qualities are not occult because the quality itself is grasped in sensation: That which is exhibited to the mind is like that which is really in the body. Gassendi's account of qualities as modes or contextures precludes any such simple solution. He speaks of his account of qualities making them more manifest, and Walter Charleton, who entitles the parallel chapter of his *Physiologia Epicuro–Gassendo–Charletoniana,* "Occult qualities made manifest," elevates that to a slogan.[17] However, it is more helpful to say that Gassendi turns *all* qualities – with the exception of size, shape, and solidity, the qualities of atoms *ex se* – into occult qualities. For we do not get into direct sensory contact with the true nature of an item in the world when we perceive sensible qualities like heat any more than when we observe occult qualities like magnetism or antipathy. The account of qualities that accompanies Gassendi's theory of perception and helps preserve the truth of the appearances thereby widens the gap between the world as it is in itself and the world as it appears to us.

[17] Charleton, *Physiologia Epicuro-Gassendo-Charletoniana*, 341–82.

10

Faith, Reason, and the Immaterial Soul

Gassendi's assessment of the extent of our knowledge of the human mind changed significantly over the course of his career. In the preface to the 1624 *Exercitationes*, he plans to show that "whatever cognition we have of intelligences and the thrice-great God is by orthodox faith," so that arguments from the natural light are in vain (3.102). He does not say anything about cognition of the immaterial human soul, but it is reasonable to suspect that the *Exercitationes* would have treated that as gained through faith alone as well. Seventeen years later, in the *Disquisitio*, Gassendi argues that we cannot conceive of an immaterial mind, although we know by faith that the mind is immaterial, and he endorses a version of the cosmological argument for God's existence. Still later, in the *Syntagma*, he provides arguments for both the existence of God and the immortality of the incorporeal human soul.

Although Gassendi is committed to the existence of God and an incorporeal, immortal human soul throughout his career, explaining knowledge of the incorporeal is always a problem for him. The imagistic tendencies of his account of mental content make it difficult to see how we can have cognitive access to incorporeal beings, and its causal tendencies make the situation even worse, given that we have no direct perceptual contact with God or angels.[1] Indeed, it is precisely this worry that lies behind the *Disquisitio*'s denial that we have an idea of the incorporeal soul. Thus, I begin by considering Gassendi's objections to Descartes's account of knowledge of the mind and the alternative he provides in

[1] I do not think that Gassendi would like to envision us as having something analogous to perceptual contact with our own incorporeal souls either, but I am less sure of this.

the *Disquisitio*, before moving on to the *Syntagma* account. Considering these two different stages of Gassendi's career helps bring out a shift in his conception of the proper relation between faith and reason. It also puts us in a position to end the book with a new account of the role of theology and religion within Gassendi's philosophy.

The Immaterial Soul in the *Disquisitio*

Interpreting the *Disquisitio* is somewhat complicated because it is an entirely critical work. This may well have allowed Gassendi more freedom of expression: He could always defend himself against charges of heterodoxy by saying that he does not endorse the views he expresses but merely intends to show that Descartes has not ruled them out. At the same time, of course, it can be difficult to discern which arguments and conclusions Gassendi is actually committed to. However, the fact that he puts forth certain arguments as objections to Descartes at least shows that he thinks they have some force. Hence it is reasonable to assume that he takes his later views on the nature of the mind to be immune to those objections.

Gassendi's main line of objection to the Cartesian account of the nature of the mind focuses on its explanatory power. He argues that, as well as raising new explanatory problems, Descartes's account of the mind fails to answer the traditional questions. For what it tells us is entirely trivial:

> When you say that you are simply a thing that thinks you mention an operation everyone was already aware of – but you say nothing about the substance carrying out this operation: what sort of substance it is, what it consists in, how it organizes itself in order to carry out its various different functions in different ways, and other issues of this sort, which we have not known about up till now. (3.300b)

Gassendi thus takes Descartes's claim that the mind is a thinking thing to simply pick out a certain entity – the unitary subject of the various cognitive powers – without telling us what the underlying substance or essence of that entity is, let alone demonstrating that it is distinct from any body. For, he argues, if Descartes had really uncovered the nature of the mind, then a number of currently unanswered questions would have clear answers:

> What sort of thing [the underlying *principium* of thought] is, how it exists, how it holds together, how it acts, whether it has certain faculties and functions, whether or not it has parts, and if it has any, what kind they are: if it does not have any and

is indivisible, how it arranges itself in so many different forms; how it performs so many functions; by what means it deals with the body; by what means it goes beyond it; how it lives without it; how it is affected by it. (3.306b)

Because Descartes has failed to answer these questions, Gassendi takes it that the claim that the mind is a thinking thing is merely a *nominal* or, perhaps, *accidental* definition – one that describes the operation or quality proprietary to the mind without making clear the essence or substance from which that operation flows.

Gassendi also argues that Descartes's account raises *new* explanatory problems and identifies two such problems: body-mind interaction and representation. The problem of body-mind interaction is a familiar one. Descartes sometimes explains interaction by positing the pineal gland as its locus, but this, Gassendi says, is unsatisfactory: There is no place – whether a part of extension or a mere point – at which material and immaterial substances could be joined with the desired result (3.405a). Thus, because bodies act by contact and contact with the immaterial is impossible, we cannot make sense of bodies affecting an immaterial mind. For bodies have only the power of motion, and motive power cannot underlie actions that do not occur in any place.

Body-mind interaction is often described as a two-way problem. However, Gassendi only seems worried about bodily causes having mental effects and not vice versa. For he does not claim to grasp what a Cartesian mind would be like well enough to say that such a mind *could not* have bodily effects; he can make only the weaker objection that Descartes does not make clear how an immaterial mind affects the body. Because Gassendi takes himself to lack any clear idea of an immaterial mind, he cannot insist that its causal powers do not suffice for action in the material world. And because he holds that God is immaterial and acts on bodies, he should not suggest that our inability to understand how the immaterial can affect the material is grounds for thinking that it cannot do so.

By raising the problem of body-mind interaction as an objection to Descartes, Gassendi commits himself to providing an account of interaction. However, he faces difficulties with interaction himself. For although he thinks we conceive of the mind as material, he also accepts that the mind is an immaterial substance and thus must strictly limit the knowledge of the mind available through reason. We know the various accidents of the mind by reflective experience, but the mind itself – the inner substance that grounds and makes possible those accidents – is hidden from sensation and reflection. We conceive of that inner substance only through an inference that construes the accidents as signs of some underlying substance. This inferential process both tells us that some substance fitted to produce the relevant accidents exists and allows us to imagine that substance, albeit by analogy with something else sensed directly. Thus, the basis of our analogy will always be a sensible, corporeal thing, so that our idea of the mind will always be the idea of a thing like some corporeal thing. Gassendi uses a subtle body like wind or ether as the basis of the analogy:

> As for the ideas of allegedly immaterial things, such as those of God and an angel and the human soul or mind, it is clear that even the ideas we have of these things are corporeal or quasi-corporeal, because (as previously mentioned) the ideas are derived from the human form and from other things that are very rarefied and simple and very hard to perceive with the senses, such as air or ether. (3.386a)

Now, it is notable that Gassendi does not claim that the idea of the mind as some sort of subtle body is a terribly *good* idea. In fact, he makes it clear that the human idea of the mind is quite a bad one. Ideas acquired on the basis of signs count as good ideas only if they have some significant explanatory power, that is, only if what the idea represents really would make possible the manifest properties and operations of the thing in question. Gassendi does not claim that his idea of the mind as a sort of subtle body actually does any significant work in explaining the manifest properties and operations of the mind. In fact, he thinks we have *no* understanding of how these properties and operations are possible.

At this point, one might wonder whether Gassendi is in any better a situation than Descartes. He attacks Descartes for claiming to know the nature of the mind when all he really has is a generic and entirely unexplanatory conception of it as some unknown thing that is not material. But although Gassendi insists that we have *some* idea of the mind, in that we can talk about it and recognize some of its accidents and operations, he denies that the idea is accurate. Isn't he just as badly off as Descartes?

The answer is no: Gassendi could defend himself on two fronts. First, he could say that at least his account of the mind makes the idea of the mind an idea we can have. We have no problem conceiving of a sort of subtle body interspersed throughout the coarser matter of the brain but, Gassendi argues, Descartes's "immaterial substance" is something we can conceive of only to the extent of saying that *it is like a material substance, only not material.* Gassendi has told us what the mind is, albeit not in any detail; Descartes has only told us what the mind is *not.* And telling us that the mind is not material gives us no help in conceiving of what an immaterial thing could be like (3.402b).[2] Second, Gassendi himself never claims to know or have an idea of the *nature* of the mind. Saying that the mind is some sort of subtle body is no more a claim to knowing the nature of the mind than saying that a magnet is composed of atoms arranged in some unknown way is a claim to know the nature of the magnet. Descartes requires an idea of the mind that is good enough to reveal the mind's nature for the real distinction argument to work: We need to know the nature of the mind in order to be certain that what underlies its modes is not a species of body. This sets the bar for the idea of the mind pretty high – it has to be such as to exhibit the nature of the mind to us, at least clearly enough to distinguish it from body. But Gassendi does not attempt to derive any dramatic ontological conclusions from *his* description of the mind, and hence his idea need not meet any such criteria. Indeed, as we are about to see, Gassendi argues that no ontological conclusions at all should be inferred from the idea of the mind.

The reader may well wonder how Gassendi could have come this close to stating that the mind is material. However, he takes pains to point out that his claim that we conceive of the mind as material is a claim about human cognition rather than an ontological claim. In matters of ontology, he follows the dictates of faith:

> I hold by Faith that the Mind is incorporeal. I hold that this issue appears too obscure by the natural light for me to claim to know the nature of the Mind . . . it is not apparent how, while the Mind dwells in the body, it can represent or understand any substance except under some corporeal *species* . . . [and thus it is not apparent] under what *species* the Mind might represent itself other than as some subtle body. (3.369a)

A number of conclusions can be drawn from the claim that we know by faith that the mind is immaterial although the mind must represent itself by means of a corporeal image and thus must represent itself as corporeal. The first such conclusion is that our idea of the mind is deeply misleading. But this seems to raise a problem of its own, in the guise of a possible conflict between the deliverances of faith and reason. Fortunately, however, Gassendi has the resources to avoid any such conflict – resources that we have already seen. For on his view, the idea of the mind as a corporeal entity is not a very good idea by the standards of reason alone. Moreover, we can see from the theory of signs and the doctrine of analogical cognition that we would think of the mind as material regardless

[2] We shall see later that this objection raises some worries about how we can conceive of God.

of whether or not it actually is material. For the explanation of why we must conceive of the mind as material makes no reference whatsoever to the actual nature of the mind, only to the nature of things evident to us through sense perception. Thus, reason itself makes us aware that our idea of the mind as corporeal is unreliable.

What position are we left in when we conclude that reason's dictates about the mind are systematically unreliable? Does this leave us with complete skepticism about the mind, or is there still a reason to continue natural philosophical investigations of the mental? Gassendi can, I suggest, maintain the latter. For he holds that empirical investigations like his investigation of the formation of ideas in the brain as a result of the physiological processes of vision give some further understanding of the mind. They do not penetrate into the nature or substance of the mind (any more than microscopic investigation of the cheese reveals its atomic structure), but they do bring about some increase in understanding. That the nature or substance of the mind is in principle concealed from reason need not make us despair of getting further knowledge of the processes underlying its manifest characteristics. We are in no worse a situation concerning knowledge of the mind than we are often held to be concerning knowledge of body.

In this vein, it is notable that the *Disquisitio* tends to be rather more skeptical than the *Syntagma* about knowledge of *body* as well. Skeptical passages in the *Disquisitio* generally emphasize the essence-appearance distinction that was also central to the *Exercitationes*:

> What we can know is this or that property of this substance or nature when it lies open to observation and is perspicuous in experience, and we do not thereby penetrate into the inner substance or nature – just as when looking at bubbling spring water, we know that the water comes from a certain source but do not thereby direct a keen look into the interior and discover that subterranean source. So, it seems, the good, all-powerful God established when he founded nature and left it to our use. For he revealed to us whatever is necessary for us to know about things by granting them properties through which we could know them and by granting us various senses to apprehend them and an interior faculty to make judgments about them. But he willed that the internal nature and, as it were, source be hidden because we do not need knowledge of it. (3.312b)

Because God has allowed us to know what we need to know, by creating things with manifest characteristics that correspond to our ability to acquire sensory information and the power to reason about those manifest characteristics, it would be unreasonable of us to complain that our cognitive faculties are not terribly well fitted for the knowledge of the hidden, underlying substance of things.

If the interpretation I have presented is correct, then the *Disquisitio*'s conceptual materialism about the mind coexists peacefully with his claim that we know by faith that the mind is immaterial. Given the limitations his theory of the cognitive faculties places on rational cognition, we have no reason to see any principled tension between the two claims and hence no reason to think that his submission to the dictates of faith is in tension with his philosophical views.

The Immaterial Soul in the *Syntagma*

In the *Syntagma*, Gassendi claims that we can have probable knowledge of the soul's immateriality and immortality by reason as well as by faith. Consider immortality. Gassendi holds that the immortality of the soul can be proven "by reasons of the sort that, even if they are not of the evidentness that the reasons of mathematics are, nevertheless attain the evidentness of physics or morality" (2.627b; compare 2.650a). Indeed, he offers both physical and moral evidence for immortality. Gassendi lays out the moral argument in the form of Patristic quotation, in this case from Chrysostomus (John of Antioch): We know that God exists and is both just and omnipotent, so that good must happen to the good and evil to the evil. However, we can see that in this life, reward and punishment are not always distributed justly. Hence there must be an afterlife in which the good are rewarded and the evil punished (2.632a). The argument from physics, my focus here, is even simpler, taking as its main premise that the rational soul is immaterial and arguing that "a thing lacking matter . . . also lacks parts into which it can be separated and dissolved, and therefore it is necessary that it perpetually perseveres as it is at any time" (2.628a). This is a common argument, but Gassendi's version is somewhat novel in its emphasis that it provides only probable knowledge of immortality. For because the indestructibility of the immaterial is only natural indestructibility, God could, if he chose, destroy any finite immaterial thing.

It is useful to look at Gassendi's shift regarding our ability to understand the immaterial soul in the context of the Fifth Lateran Council's 1513 request that Christian philosophers attempt to demonstrate that the immortality of the soul can be known by natural reason.[3] As a result of

[3] Evidence that this injunction was still widely recognized in Gassendi's time is provided by Descartes's remark that "in its eighth session the Lateran Council held under Leo X

this, we find seventeenth-century textbooks like Eustachius' saying that even though it is well known from the Bible, the Church Fathers and various Councils that the soul is spiritual and immortal, nevertheless it is desirable to demonstrate this by reason as well.[4] Eustachius adopts the common tactic of demonstrating immortality from immateriality, and argues, among other things, that the rational soul must be immaterial because it perceives "common natures" abstracted from particular matter and incorporeal beings like God and the separated soul.[5] Gassendi is well aware of the Lateran Council's decision (2.627a), and he endorses the argument that perception of common natures proves immateriality as well as the argument from immateriality to immortality.

Eustachius describes the immaterial soul as an informing form that is "intimately and substantially united" with the rest of the person and together with it constitutes a corporeal substance that is "per se one."[6] For Gassendi, the rational soul is an "incorporeal substance" (2.440a) "infused into the body... like an informing form" (2.440a); later in the same chapter, it is a "substantial form" (2.466a). Concerning human knowledge of the immaterial soul, Eustachius writes that

> It is very probable that in this life the soul does not understand itself through its essence, nor through an immaterial species of itself, but by the intervention of some intelligible species that it abstracts from the senses, and [that it does not understand itself] except by a reflex act and by some interposed discourse, by means of which it derives *notitia* in itself from the perception of other things.[7]

Eustachius is here appealing to a Thomistic solution of a difficulty within Aristotelian accounts of cognition. The difficulty is this. If there is nothing in the intellect that was not previously in sense – that is, no intelligible species without a sensible species – then how can insensible entities like

condemned those who take [the position that the immortality of the soul is known only by faith] and expressly ordered Christian philosophers to refute their arguments and use all their powers to establish the truth" (*Oeuvres de Descartes*, 7.3).

4 It is worth noting, however, that not everyone in the period accepted the challenge: There were Scotist claims that immortality could not be demonstrated by strictly philosophical means, and Toletus suggested that Pomponazzi might have been right in thinking that natural philosophy cannot demonstrate immortality. See Kessler, "The Intellective Soul," 509 ff. The worry is not, however, Gassendi's worry that natural philosophy cannot even demonstrate immateriality; these writers accepted that the soul was a form, not a body.

5 Eustachius, *Summa philosophiae quadripartita*, 266 (Section 3.4.2).

6 Ibid., 267 (Section 3.4.3). The relevant contrast here is with an assisting form, which does not constitute a unity with that which it informs. Gassendi also recognizes the distinction: Our souls are in our bodies as informing forms, but when angels move bodies they are in those bodies as assisting forms (2.444b).

7 Ibid., 273 (Section 3.4.5).

the immaterial soul be cognized? He answers this question in terms of intelligible species and *verbi mentis* (mental words) or *conceptus* (conceptions) where an intelligible species subsists in the mind even when not actively contemplated and a *verbum* is an act. Aquinas held that we can have a *verbum* or occurrent thought of some *x* without an intelligible species of *x*, so long as there is an intelligible species of some *y* analogous to *x* and we have discursively constructed some specification of the respect in which *x* and *y* are analogous.[8] The Thomistic view lies in the background of Gassendi's account of how we construct *notitia* of unsensed things, and indeed a lot of what Gassendi has to say about the incorporeal soul is simply taken over from scholastic accounts.

Gassendi's probable arguments for the immateriality of the rational soul rely on the existence of certain mental operations that cannot in principle be performed by a material thing.[9] First, the mind or intellect can understand itself and its functions, while no material thing can act on itself directly (2.441a, 2.451b). Gassendi here runs together the mind's capacity to reflect on itself directly, a capacity that he had explicitly denied in the *Disquisitio*, with its capacity to notice its functions, a capacity that he had there allowed but did not think required incorporeality. Second, the mind can apprehend certain notions lacking imagistic content, such as the notion of an immaterial substance, formed by way of analogy and negation (2.440b). Third, the intellect can have some grasp of universals although genuine universals cannot be apprehended by sense or derived from sensory apprehension (2.441b, 2.451a).

None of these arguments are at all novel, and it is interesting to note that Gassendi often ran them in reverse in his earlier work, arguing, for instance, that we cannot entertain genuine universals because we lack the immaterial intellect needed to do so. Indeed, it is difficult to know how seriously to take these three arguments, especially the argument from cognition of universals. For Gassendi does not appeal to grasp of genuine universals elsewhere in the *Syntagma*. The *Institutio logica*, for instance, explains how universal cognition is derived from sensory ideas by a process of abstraction and generalization, without any need for genuine universals.

What about the claim that the intellect can reflect on itself directly? Although this appears to imply that we can form a positive conception

[8] For this account in general, see Aquinas, *Summa theologiae*, 1 q.87 and 1 q.88. For mental words in particular, see 1 q.93 a1.

[9] These arguments are discussed in some detail by Michael and Michael, "Gassendi on Sensation and Reflection," 585 ff.

of our own minds, Gassendi does not offer any account of what such a conception consists in. Instead, he seems to still maintain that the immaterial mind is known only under the false guise of a subtle body or by way of negation. Our idea of the incorporeal mind represents some unknown thing that is different in kind from the corporeal things we experience because it is incorporeal and can perform these three operations. This idea is barely more informative than the earlier idea of the mind as a corporeal subtle body. Certainly, it gives us no more resources for further investigation of the mind. Indeed, it is remarkable that when Gassendi adopts the claim that we know by reason that the mind is immaterial, nothing *else* in his account of the mind changes. The claim that the mind is incorporeal neither rules out the account of impression-formation in the brain nor provides us with any new explanation of the states and operations of the mind. Gassendi's switch in views brings together the ontologies of faith and natural philosophy without making any real impact at all on natural-philosophical accounts of the mind.

Now in one sense, we should have expected this. Recall Gassendi's objection to Descartes that postulating a simple, indivisible, immaterial mind rules out any explanatory resources. Although Gassendi does not explicitly assert that the incorporeal mind is simple and indivisible, he does not think that our idea of it depicts it as having any parts or structure, and hence it is no more useful for explanatory purposes than an idea representing a simple, indivisible, incorporeal thing. Thus on Gassendi's view, whether or not we have an idea of the mind that is in keeping with the dictate of faith that the mind is immaterial, the immaterial mind whose existence we accept has very little explanatory role to play in the natural philosophy of the mind. I take this to be a straightforward expression of Gassendi's commitment to discovering the limits of human knowledge, and although it threatens the plausibility of his rational arguments for the immateriality and immortality of the soul, there is no reason to think it threatens the more fundamental claim that we know by faith that the mind is immortal.

The third argument, namely that the mind can apprehend certain notions devoid of imagistic content that no material system could apprehend, is somewhat more plausible and central. For Gassendi is committed to it being possible for us to cognize God and the immaterial mind without corresponding images of them. Even were he to think we know the immateriality of the soul only by faith, he would still need us to have some cognitive access to that thing whose immateriality we know by faith alone. More generally, Gassendi needs some form of analogical cognition to explain the possibility of cognition of atoms, molecules, and underlying corpuscular structures too small to see. So this third argument is not ad hoc in the way the first two are. Despite the apparent importance of non-imagistic cognition, however, it receives little development. Thus, it is hard to determine why Gassendi thought only an immaterial mind could perform the necessary discursive acts, if indeed he was committed to this view for reasons beyond the obvious pragmatic ones.

The Immaterial Soul and the Body

The interaction of mind and body is perhaps the most famous of the traditionally studied problems of seventeenth-century philosophy, and it is well known that Gassendi, like others, objected that Descartes's ontology made it impossible to understand how the body can affect the mind. Gassendi does not face this problem nearly so squarely within his own theory. He suggests an explanation of the causal connection between immaterial soul and body in terms of the mediation of the corporeal soul, writing that "the mind is not judged to be joined immediately to the coarse body, but to be united first with the sensitive and vegetative Souls" (2.444a). This is reminiscent of neo-Platonist accounts on which spirit serves as an intermediary between body and soul, but Gassendi does not really have the resources within his own system to coopt that solution.

Gassendi recognizes that there is indeed a problem explaining "how the Mind, or Rational Soul, could be said to be united with a body, whether it is coarse or subtle" (2.444a). For even the subtle sensitive soul is "infinitely distant from the incorporeal" (2.444a). The natural view, Gassendi suggests, is that unification takes place through some kind of "glue" (2.444a), a term that he elsewhere uses loosely enough to encompass the hooks that stick atoms of certain bodies together. But this cannot work for the unification of mind and body. Instead, Gassendi suggests a tripartite ontology: "the author of nature supplied three kinds of natures, the first of purely incorporeal things or angels; the second of purely corporeal things or bodies; and the third of the agreement [*constantium*] of corporeal and incorporeal, or men" (2.444a). Thus, the condition of the rational soul is such "that it has by its own fitting nature a destination and inclination toward body and the sensitive soul" (2.444a). The explication that follows, however, makes clear that what Gassendi has in mind is a view of body as matter and rational soul as its *actus* – a view hardly consonant with what he says elsewhere in the *Syntagma*.

It is helpful to divide the interaction issue into two distinct subissues: how the mind affects the body and how it is affected by it. In his account of agency, Gassendi writes that

> the human soul, which is the intellect or mind, and therefore incorporeal, does not elicit actions except intellectual or mental and incorporeal ones . . . the soul that senses, grows, and is endowed with a power of moving the body, and is therefore corporeal, is what elicits corporeal actions. (1.334b)

This claim relies on a general principle that Gassendi here, although by no means everywhere, grants, namely that (with the exception of God) an incorporeal thing cannot elicit corporeal actions. Gassendi thinks this general principle causes problems for angelic action. We do not want to say that angels are corporeal because (bracketing out some issues about how the fires of hell act on angels) faith dictates that angels are incorporeal. But angels aid in God's special providence by acting on the human imagination. How is this possible, if the incorporeal cannot act on the corporeal? Gassendi explains that "God imbued the angels destined for these duties with a power that is beyond the order of nature and (because there is nothing it cannot do) special and unknown to us" (1.335a). This hardly helps explain interaction between human minds and human bodies. What is worse, Gassendi grants that the action of incorporeal angels on corporeal human beings is not open to explanation by the natural light: "If we hold as certain by Faith alone that Angels are incorporeal and act as we read they do in the Holy Books, it seems that by this reason we can hold something that physical reasoning [*Physica ratio*] does not fully grasp. And yet this state of affairs is permitted by Sacred Theology" (1.335a). Is he also willing to grant that physical reasoning cannot grasp the way the incorporeal soul affects and is affected by the body? He writes: "We should not disproportionately linger on the fact that the Intellect, an incorporeal faculty, immediately uses Phantasms, corporeal species . . . it is commonly admitted that the incorporeal soul is immediately joined to the body and uses corporeal limbs to move" (2.449a; compare 1.99a).[10] The fact that the incorporeal intellect constitutes a unity together with the body is supposed to make it possible for it to use corporeal species. Thus, even if we cannot understand *how* interaction occurs, we should not let

[10] For "ut Intellectus, facultas incorporea, Phantasmate, species incorporea, immediate utatur," I read "Phantasmate, specie *corporea*."

this stop us because *every* account of the soul is going to have this problem. After all, Gassendi identifies "what is stipulated by the Holy Faith" as that "the Mind . . . is an incorporeal substance," so he can safely assume that his audience will face a problem of interaction no matter what the rest of their theory looks like. At this point, it is important to remember that Gassendi objected to Descartes not only because he had *failed* to explain interaction and union but also because he had made it *impossible* to explain them.[11]

Gassendi's account of the immaterial is less than satisfying. Indeed, because the *Syntagma* suggests that we can know the soul's immateriality and immortality by reason as well as by faith while giving no good arguments, it is much less satisfying than the *Disquisitio.* Gassendi is almost entirely relying on faith to provide an answer that reason can pretend to support: Neither the claim of self-reflexive ability nor the claim of universality can really be developed in accordance with the theory of cognition Gassendi puts forth elsewhere.

However, it is important to notice that this is no implicit adherence to a theory of double truth, as it would be if reason demonstrated that the soul is corporeal rather than simply failing to demonstrate that it is incorporeal. Such a failure to demonstrate incorporeality would seem hard to accept in a system purporting to give a complete, rational explanation of the universe – but an insistence on the explanatory limits of natural philosophy is one of the dominant themes of Gassendi's work.

Knowledge of God

Although the Epicureans accepted the existence of gods who had no concern for human affairs, their views were often branded as atheist in a seventeenth-century context where to believe in the wrong god or gods was to be an atheist. Thus, one crucial part of Gassendi's rehabilitation of Epicurean atomism is to demonstrate that God exists and to convince his readers that God plays a crucial role in Gassendi's philosophical system.

Gassendi's demonstration of the existence of God is, like many of his arguments, at once a proof of the existence of something and an explanation of how we acquire the idea of that thing. The *Syntagma* provides two

[11] I do not mean to suggest that Gassendi takes interaction and union to be the *same* issue. Rather, the assumption is that once we understand the union then we will understand interaction because there is no great difficulty understanding how parts of a whole reciprocally affect one another.

arguments for the existence of God, the argument from universal anticipation, which claims that men "have some *notitia* of God impressed by nature itself" and the argument from the spectacle of nature (1.290a–b). Gassendi does not claim that *all* men have an idea of God, much less the *same* idea; indeed, in the *Disquisitio* he brought up cases of men who lack the correct idea of God in objection to Cartesian nativism. However, he argues, the fact that some unfortunate men lack an idea of God is entirely consistent with its being natural, just as the unfortunate fact that some men are born deformed or blind is entirely consistent with the naturalness of possessing limbs and a faculty of sight (1.290b). For a natural tendency can fail to produce what it normally produces if conditions are not right. Seeds, for instance, have a natural tendency to produce adult organisms, but all sorts of factors can interfere with their germination and growth, or influence the mature organism to develop more or less fully. To say that we have a natural idea of God is simply to say that we have a tendency to acquire that idea in the right conditions.

In what conditions do we acquire this natural idea? Gassendi recognizes that most people acquire it by testimony, and that it can also be acquired by revelation – but the most interesting way in which we acquire the idea of God is by inferring to the necessity of a cause of order in the world, as the argument from the spectacle of nature describes. Indeed, this argument is of central importance in Gassendi's system, and its centrality is signified by its privileged location in the *Proem* to the *Syntagma*. After examining and refuting Epicurus' claim that the goal of natural philosophy is to dispel religious superstition, Gassendi claims that the true goal of natural philosophy is to point us toward the divine:

> *Physiologia* is the [subject] that, by contemplating the nature of things in general, draws the conclusion from the magnitude, variety, disposition, beauty and harmony of its wonders that there is a most wise, powerful, and good God [*Numen*] by whom [the universe] is governed... and therefore natural reason leads us to recognize that the excellence and beneficence of this God should be shown a reverence that is religion itself. (1.128b)

In a similar vein, Gassendi endorses what he calls the Epicurean categorization of theology against the Platonic, writing to Louis de Valois in 1642 that even though Plato considered theology a distinct branch of philosophy from physics, "the Stoics and also Epicureans made [theology] a part of physics, as physics, investigating the causes of things, recognizes in them divine causes, especially the productive, conserving and governing cause" (6.137b). Gassendi's rhetoric enlists all of natural philosophy in the service of knowledge of God. La Poterie, Gassendi's secretary and later biographer, also uses this rhetoric, writing that even after Gassendi stopped giving public sermons for health reasons, "he still applied himself to instructing the people, not by his voice but by his pen, and to inciting knowledge of the true God through knowledge of natural things."[12]

But what is the idea of God acquired through this sort of reasoning, and how does reasoning give us the idea of an incorporeal God when all ideas come from or through the senses? For Gassendi – as for Aquinas – our knowledge of God is gained through an analogy that we know is misleading, just as we know that the analogy that yields an idea of the incorporeal mind is misleading. In the *Syntagma*'s crucial chapter "On the form under which we apprehend God," Gassendi attempts to reconcile the Thomistic view with the Epicurean claim that the gods have human form by interpreting the Epicurean claim as concerning our idea of God and not God's real nature. It is hard to believe that this is anything but a deliberate misreading; Gassendi was not, in general, a careless reader.

Gassendi argues that our knowledge of God is by analogy mainly by arguing against what would seem to be the only other option open to him, the *via negativa*. He writes that "the species [of God] cannot be a mere negation but is something positive" (1.296a), for if the idea of God were merely the idea of something ineffable and indefinable, then we would not be able to think about him at all. Rather, "nothing is more human than that, when thinking about God, we imagine a human form" (1.296a). For instance, we conceive of God as perfectly intelligent by starting from our own limited understanding, and so forth. Similarly, we conceive of his incorporeality by ascending through different degrees of relative lack of corporeality: Water is incorporeal with respect to earth, and air with respect to water, although they are both bodies.

Gassendi recognizes a possible objection: If you keep removing degrees of corporeality in thought, you end up with nothing. However, he replies, such objectors "do not recognize the fallacy by which they confuse imagination or, as I might say, intuitive understanding with reason, or inferential understanding [*consequutiva intelligentia*]" (1.297b). Of course, intuitive understanding – which for Gassendi is primarily the sort

[12] La Poterie, "Memoires," 235.

of understanding deriving from literal intuition, visual perception – is primary:

> But beyond this [intuitive] understanding there is in us something else, by which we perceive something not intuitively but inferentially. And hence we do not so much perceive what it is (since we do not intuit it) but rather as it were suspect what it is and judge from the necessity of an argument what it must be. (1.297b)

Thus, our idea of God is not intuitive but inferential; this is the crucial Thomistic step, and the one at which Gassendi reintroduces the claim that an incorporeal mind is necessary in order to have inferential understanding (2.451a).

The product of such inferential understanding – which produces a thought that does not correspond precisely to any underlying disposition (2.44b) – is variously called an idea, cognition, or *notitia.* Here Gassendi brings out the "rude example" of the masked man from the *Meditations.* When we see a masked man, we intuit the mask and not the face, but "we infer with necessity that some face is hidden beneath the mask" (1.298a). The example is somewhat misleading because it combines two distinct steps. First we form an idea or *notitia* of God by successively removing degrees of corporeality; then we come to believe that such a God must exist as the wise ordering cause of the universe.

Inferential understanding cannot present its object with any evidentness, and the idea of God arrived at by it is not supposed to be accurate: "God, *qualis in se est,* cannot be represented by the mind. Certainly, we cannot cognize the perfections of God except by some analogy or comparison to those perfections that have been observed in other things" (1.302b-3a). Of course, Gassendi allows that the perfections and attributes we understand analogically are the typical ones. God is immense, omniscient, eternal – by which Gassendi means sempiternal – omnipotent, possessed of intelligence and will, and unique (1.303a). He is pure act (1.306a) and possesses unlimited *scientia,* indeed the only *scientia* anyone possesses (1.207b). He is not restricted by preexisting possibilities and can do anything but make both halves of a contradiction simultaneously true, and this is not a limitation in God's nature, for the contradiction is *ex parte rei* (1.308b). For God constituted the laws of nature and can break them (as he does, for instance, in transubstantiation) (1.308b, 1.381a). God knows all the unactualized possibilities, past, present, and future (1.307b). God created the world *ex nihilo,* although the Epicurean maxim *ex nihilo nihil fit* applies within the order of nature (1.234a). And in addition to creating the world, God conserves it in existence (1.323b, 1.485b, 3.346a) and concurs with the genuine causality of secondary causes (1.326a, 1.337b).

Immensity and sempiternity stand out as the problematic items in Gassendi's list of divine perfections. We saw in Chapter 5 that Gassendi needs to make the uncreated nature of a "real being" like space acceptable within a Christian context, and that he is at least tempted to accomplish this by holding that space is God's immensity. However, Gassendi must in the end think that only the misleading, analogical idea of God implies that space is God's immensity: "the Immensity of Infinitude of God is conceived by us by comparison to place, which is itself also immense in its way, but nevertheless it must be understood that the divine Substance . . . is of a kind of immensity without relation to place" (1.304b). For "even if God is said to be Everywhere, or in every place, nevertheless it would more properly be said that he is *in se*, since that, as it were, diffusion by which he is present in every place is only a certain consequence of that, as it were, diffusion that he previously has in himself" (1.304b). However, Gassendi is not entirely consistent about whether God's spatiality is analogical or real, for in the same texts where he qualifies the claim that God is in every place, he is also willing to make the suggestion that God can act on the world in virtue of being everywhere. It may be this suggestion that led Gassendi's followers to the heterodox suggestion that God is literally, rather than analogically, a body. For instance, Thomas Stanley writes that

> neither is [God] therefore a gross Body, no not the most subtle that can be coagmentated of Atoms; but he is altogether a Body of his own Kind, which indeed is not seen by Sense, but by the Mind; nor is he of a certain Solidity, not composed of Number; but consists of Images perceived by Comparison; and which, compar'd with those that ordinarily occur, and are called Bodies, may be said . . . to be (not Body, but) as before I said, Resemblance of Body; and (for Example) not to have Blood, but a certain Resemblance of Blood. . . . In the mean time, I must intimate by the way, that he is not such a Kind of Body as is coagmentated of Atoms; for then he could not be sempiternal.[13]

The claim is confused, but at any rate Gassendi is not willing to go so far in explaining cognition of God on analogy with cognition of bodies. Indeed, he insists that when used in connection with God, the term 'incorporeal' "does not signify the bare negation of body and corporeal dimension, but also brings in a true, genuine substance and a true, genuine nature with appropriate faculties and actions" (1.183b).

[13] Stanley, *The History of Philosophy*, 630b.

This is in line with the Thomistic claim that "we are unable to apprehend [the divine substance] by knowing *what it is,*" but that at the same time "we are able to have some kind of knowledge of it by knowing *what it is not.*"[14] However, Gassendi has no real way to specify what incorporeality involves. In this respect, his accounts of knowledge of God and knowledge of the human soul are very similar. In both cases, he is careful to avoid implying any contradiction between the deliverances of faith and reason but makes only a half-hearted attempt to explain how it is that reason arrives at the truths dictated by faith.

Faith and Reason

Gassendi needs to explain cognition of God and the immaterial human soul in order for his natural philosophy to be rhetorically and doctrinally acceptable to his intended audience. Epicureanism had faced accusations of atheism and materialism, and Gassendi was concerned to rebut these charges. His contemporaries and close successors did not express much concern with the success or failure of that rebuttal. Arnauld worried about the "damnable" consequences of Gassendism, but these worries are not fleshed out, and there is no reason beyond the choice of words to think he has theological consequences in mind.[15] Morin, as we have seen, launched an attack on Gassendism that started with Copernicanism but came to encompass all the fundamentals of atomism.[16] Beyond these two cases, however, Gassendi and his Epicurean atomism suffered no significant theological critiques, and his works – unlike those of, say, Malebranche or Descartes – were never placed on the Index.[17]

However, some twentieth-century scholars have argued that Gassendi should be characterized as a libertine or that his system contains an implicit tendency toward materialism and atheism. The most prominent argument of this sort is put forth by Olivier Bloch in his *La philosophie de Gassendi.* Bloch takes there to be a serious and inescapable tension between Gassendi's explicit statements about God, providence, creation, and the human soul on the one hand, and the fundamentals of his natural philosophy on the other. In particular, he describes tensions

[14] Aquinas, *Summa Contra Gentiles,* 1.14.2.

[15] See Arnauld, *On True and False Ideas,* 60.

[16] Objections focusing on atomism rather than Copernicanism take center stage in Morin's *Réponse à une longue lettre* and *Defensio suae dissertationis,* the works to which Gassendi's *Anatomia ridicule muris* and *Favilla ridicule muris* reply.

[17] For details of the case of Descartes, see Ariew, *Descartes and the Last Scholastics,* 173–81.

between the arguments for the existence of the immaterial soul at *Syntagma* 2.3a.10 and the roughly materialist account of cognition provided earlier; between Gassendi's sympathetic treatment of the Epicurean principle *ex nihilo nihil fit* and his belief in creation *ex nihilo*; between his claims that matter is active and that God is the only original cause; between his Galilean account of motion and his overt geocentrism; between his account of space and time as uncreated and the doctrine that God is the author of all things; and between orthodoxy and his views on the plurality of worlds, Lucretian evolution, and the limited extent to which we can have knowledge of God.[18]

Bloch's argument is subtle and attractive, although, I shall argue, ultimately unsustainable. It is important to recognize that his claim is not that Gassendi's professions of orthodoxy are insincere or disingenuous. He writes that Gassendi personally "was no doubt a believer" but holds that his philosophy nevertheless contains a strong orientation toward materialism. Thus, he refers to his study as "the anatomy of an abortive system."[19] Gassendi's own religious beliefs and his indisputable intention to make Epicureanism acceptable to a Christian audience, emphasized by Margaret Osler, are thus irrelevant to assessment of the truth of Bloch's claim.

Bloch is well aware of the sort of documentation of Gassendi's piety provided by his contemporaries and of Gassendi's concern to point out the dangerous theological errors of writers such as Fludd. However, he claims that Gassendi's natural philosophy provides a framework that, if fully systematized, would leave no space for theological metaphysics. By characterizing Gassendi as an ideologue, Bloch draws a sharp distinction between the kinds of reasons Gassendi had for accepting the majority of his natural philosophical beliefs and those he had for accepting the existence of God, providence, and the incorporeal human soul.

Bloch delineates two distinct stages in Gassendi's attempt to integrate Epicureanism and theological metaphysics. At the first, pre-1641 stage,

[18] The most important are the cases of God and the human soul. I think it is reasonably clear that there is something more than tension in the heliocentrism case: Even though Gassendi may have submitted to the decrees of the church, I do not think he ever really believed those decrees were legitimate. However, the Galileo case does not have many ramifications, and in any case it hardly differentiates Gassendi from his peers. The other cases, I argue, do not provide evidence of any genuine incompatibility between faith and reason. Reason may suggest that some claims contrary to faith are probable. But conflict between faith and mere probability is no problem: Even Thomas allowed such conflicts. See, for instance, my discussion of Lucretian evolution in Chapter 8.

[19] Bloch, *La philosophie de Gassendi*, 328.

he simply juxtaposes an Epicurean philosophy of nature with various orthodox affirmations, providing no commentary on how they relate. The second stage begins with the *Fifth Objections* and evolves in the *Disquisitio.*[20] At this stage, Bloch thinks that Gassendi is trying to integrate his materialist philosophy into an all-encompassing religious world view and thus to arrest Epicureanism's ineluctable trend toward materialism. On Bloch's account, Gassendi's main tool in doing this is the doctrine of double truth, as it was resurrected after the condemnation of Pomponazzi, and he understands Gassendi's various remarks on what is known through faith and *ratio philosophandi* as expressions of such a doctrine. [21]

An additional clue is provided by Carla Rita Palmerino, who points out one important difference between the *Proem* to the *Syntagma* and its antecedent, the 1634 preface to the unpublished *De Vita et Doctrina Epicuri.*[22] The *Syntagma*'s *Proem* ends with Gassendi's insistence that "[t]he name of no sect applies to me . . . except only the orthodox, that is, those who accept the greatest, Catholic, Apostolic, Roman Religion, to which I adhere uniquely. . . . I conduct myself so that I always prefer the reason of [its] authority" (1.29b–30a). This statement has no analogue in the *De Vita et Doctrina* preface. Palmerino construes the absence of any such statement as follows:

> [W]hether [the *Syntagma*'s conclusion] really reflects the end-point of Gassendi's own philosophy or the intellectual development it had undergone since 1634 is doubtful. Much rather, this attached profession of orthodoxy seems to mark the difference between a privately circulated manuscript and the public and published pronouncements of the Prevost of Digne who could not speak *De philosophia universe* without invoking the Universal Church.[23]

This is a natural reading. But the fact that a profession of orthodoxy is included in a published work but not in one circulated to friends could be taken to indicate that the profession is not *necessary* in an only semipublic context just as well as it could indicate that the public profession is insincere. Even if Gassendi had no worries about the compatibility of a scholarly account of Epicureanism with religious orthodoxy, he might

[20] The delineation of these two stages is a bit complicated: Before 1641, Gassendi had published no positive accounts of his natural philosophy – only the astronomical work that established his reputation and the occasional tracts against Fludd, Herbert of Cherbury, and the Aristotelians. Bloch's view is based on an examination of manuscripts that, although more or less complete, were never considered finished.

[21] Ibid., 476–81.

[22] Palmerino, "Pierre Gassendi's *De Philosophia Epicuri Universi* Rediscovered."

[23] Ibid., 294.

still have wanted to ensure that foolish or weak-minded readers did not get the wrong idea. Moreover, the Epicurus project began as historiography and only gradually transformed itself into a positive philosophical program.

Of course, Gassendi's choice of Epicureanism as his subject was always at least partly motivated by philosophical considerations. As Palmerino points out, Gassendi argued for the superiority of Epicureanism over the other ancient Greek systems beginning in the early 1630s. But because he presents the choice as one between competing pagan systems that all need amendment, an argument for the superiority of Epicureanism does not imply commitment to the truth or completeness of the Epicurean system – merely a commitment that a suitably revised form of Epicurean atomism can be reconciled with orthodoxy at least as well as any of its competitors.

I think that Bloch overstates the extent to which the claims he identifies as affirmations of orthodoxy conflict with those he regards as scientific, and I have a suggestion as to why. One main concern of Bloch's work is to downplay the significance of epistemology in Gassendi's work. This may serve as a useful corrective to characterizations of Gassendi as the empiricist predecessor of Locke, but it also magnifies the appearance of tension between orthodoxy and natural philosophy. Consider the case of the immortality of the soul, where Gassendi articulates two theses that are prima facie in tension: that we cannot avoid conceiving of the human mind as a subtle body, and that the mind is incorporeal. I have argued that these two theses are entirely consonant, given that Gassendi argues that we have good reasons to think that our conception of the human soul is inadequate – reasons pertaining to the limitations of the human cognitive capacity. Paying careful attention to the epistemic status Gassendi attributes to various aspects of his system dissolves the appearance of contradiction.

The plausibility of Bloch's case and in particular his claim that Gassendi has recourse to a theory of double truth depends on whether there really are contradictions within Gassendi's philosophy or between it and the religious and theological doctrines he accepts. I argue that cases like the case of the immortal soul are *not* ones where reason tells us one thing with demonstrative certainty and faith another. Rather, they are cases where *probability* tells us one thing and faith another – and in such cases, we can reject the conclusion of probable reasoning without making any overarching judgment that reason is to be devalued or that it is in conflict with faith. I do not see any reasons to think such a move is ideological that would not also commit one to thinking of the bulk of seventeenth-century philosophical discourse concerning matters relevant to Christianity as ideological.[24]

It is helpful to contrast Gassendi's use of probability in this case with the way he introduces epistemic concerns when discussing heliocentrism, where epistemology serves a useful rhetorical function but is not, I think, really doing any work. Gassendi uses the Epicurean account of hypotheses – on which two inconsistent hypotheses can both count as true in virtue of both conforming to the appearances – in order to avoid denying either geocentrism or Copernicanism. Now, this is a well-attested feature of Epicureanism, and Gassendi uses it in other cases. For instance, he suggests it as a way of dealing with the plausibility of two hypotheses concerning the original source of seeds: that God created certain first seeds that replicate themselves, and that God created dispersed atoms that arranged themselves into seeds through their motive power. The first hypothesis accords with Biblical literalism, while the second is a creationist version of Lucretian evolution, one that replaces chance with divine ordination.

The creation of seeds is, like heliocentrism, a theologically loaded example, but one could also adumbrate cases where two different corpuscular mechanisms have roughly equal probability: Gassendi sometimes accepts as probable two competing hypotheses that accord with the appearances in cases where there is no theological motive. Although in the Galileo case it is hard to avoid seeing a kind of desperation in Gassendi's use of the Epicurean move, probabilism permeates the *Syntagma* so deeply that it is not plausible that his recourse to it is simply ad hoc. Almost all Gassendi's natural philosophical claims are supposed to be probable rather than demonstrative. Thus, although probabilism allows Gassendi to reconcile faith and reason at a number of points, it is unlikely that it is inserted into his system simply for this purpose.[25]

The fact that Gassendi's arguments are often supposed to carry only probability is important for understanding how we should deal with the

[24] I am not entirely sure that this would worry Bloch. Perhaps he would be sympathetic to the claim that basically all seventeenth-century religious discourse – or all religious discourse in general – is ideologically rather than philosophically motivated. However, if this is so, then his analysis is not telling us anything about Gassendi in particular, and thus would lose a great deal of its interest.

[25] Moreover, probabilism is apparent as early on as the *Exercitationes*, where Gassendi took the strongly fideistic line that all cognition of God and separated substances is by faith alone.

tensions Bloch notes. It is helpful to look at these cases anyway, for because Gassendi never articulates any general view about the relation between faith and reason, we must elicit a view from particular cases. Faith or "sacred doctrine," for Gassendi and his Catholic contemporaries, is the set of propositions derived from the Bible, the Church Fathers, and the decrees of the Pope and various Councils. Individual propositions, articles of faith, can be related to the deliverances of reason in one of three ways. First, some articles of faith – such as the existence of God and the immateriality of the soul – can also be demonstrated by philosophical reasoning. Second, there are some articles of faith that reason renders probable without being able to prove demonstratively, such as the finite spatial extent of the world. Finally, there are some articles of faith that reason renders *improbable*. Examples include the temporal end of the world and – according to the *Disquisitio* but not the *Syntagma* – the immateriality of the human soul.

The existence of propositions falling into the first two categories clearly does not ground any conflict between faith and reason. Because Gassendi allows that good probable reasoning can have false conclusions, the existence of articles of faith rendered improbable by reasoning does not provide evidence of any genuine conflict either. Of course, it would be a problem for reason to *demonstrate* the falsity of articles of faith – but Gassendi need not and does not admit any such demonstrations.

Other propositions are more or less constitutive of orthodoxy without being articles of faith. One example is the denial of the plurality of worlds. Both Epicurus and Bruno argued in favor of the plurality of worlds. Against this, Gassendi holds the uniqueness of the world to be "in the greatest consonance with the principles of sacred faith and religion" (1.337a). The uniqueness of the world cannot be demonstrated because God *could* create other worlds. However, God has only given us knowledge of this world, and hence it is improbable that there are others. Aside from this fact, probability tells us nothing about the uniqueness or plurality of worlds, and thus we ought to reject the plurality of worlds as improbable. Another example is the nature of creation. Gassendi tells us that Biblical accounts of creation are incompatible with his modified version of Lucretian evolution if they are read literally but is unwilling to insist that a literal reading of the claim that God created the world in six days is necessary. He devotes some attention to showing that atomism and the theory of seeds are compatible with a literal reading of the Bible but remains silent on whether there are any other reasons for denying a modified version of Lucretian evolution. Thus, he delivers no verdict on whether probable reasoning supports or contests a six-day creation. In both of these examples, the relation between what reason suggests and what Gassendi takes faith to demand is complex, and reason does not simply bolster faith. However, I see no trace of a doctrine of double truth. Gassendi never allows demonstrative reasoning to contradict faith.

Bloch's reading of Gassendi has been challenged in recent English-language scholarship, most prominently by Margaret Osler.[26] Osler insists on the sincerity of Gassendi's claims that God created the world, rules it through his providence, endowed men with immortal souls, and so on.[27] She takes these claims to support her thesis that there are deep theological motivations, in particular voluntarist motivations, underlying Gassendi's natural philosophy. Osler characterizes her argument as opposed to Bloch's, but the two views are actually compatible. One could perfectly well hold both that Gassendi was motivated by a particular theological doctrine and that he failed to arrive at a natural philosophy consistent with it.

Osler distinguishes between two styles of science: one exemplified by More, Huygens, and Descartes, and the other by Gassendi, Boyle, Locke, and Newton, who "emphasize observational and experimental methods," think their conclusions "achieve at best some degree of probability," and allow that "some [events] lie beyond the reach of scientific theory."[28] She associates this style of science with voluntarism, understood as comprising three theses. First, there are no necessary or eternal truths independent of God; the nature of the created world depends only on God's will. Second, regularities in the course of nature are contingent and maintained freely

[26] Brundell, *Pierre Gassendi*, also argues against Bloch, and in a manner I am much more sympathetic to. He argues that although Gassendi made some obvious changes to accommodate Epicureanism and Christianity, "baptizing" Epicureanism – to use Osler's resonant phrase – was never a major concern. Although I agree with Brundell on this point, we differ as to the nature of Gassendi's chief philosophical motivation. Brundell thinks of Gassendi's dominant project as the negative one of combating Aristotelianism. I suspect that Gassendi may well have taken it for granted that Aristotelianism would soon be replaced by a new philosophy: The question in his mind was what that new philosophy would be. Gassendi's natural philosophy was put forward for its own sake.

[27] Osler, *Divine Will and the Mechanical Philosophy*, 56–77.

[28] Ibid., 223. Osler's characterization of the Gassendi–Boyle–Newton style of science is not entirely accurate so far as Gassendi goes. For one thing, Gassendi *does* think that certain natural philosophical claims – like the existence of atoms and the void – are known with demonstrative certainty. For another, although Gassendi relies heavily on a rhetoric of observation and experience, it is not clear how much observation and experience actually contributed to his results. Most of his embryology, for instance, simply relies on his predecessors' observations.

by God. Third, God can literally do anything at any time (save make both parts of a contradiction true at once), so there are no a priori truths, and we can have no rational insight into causal relations.

I am not entirely certain what Osler proposes as evidence for the claim that Gassendi's views are in large part motivated by theological voluntarism. She documents the apparent sincerity of Gassendi's commitment to the doctrines of the Church carefully, but this does not help her case against Bloch. Moreover, even though it is correct to call Gassendi a voluntarist in her sense of the term, there is not as much text talking about voluntarism as one might expect if it is as significant as Osler suggests. Although on Osler's reading, the distinction between God's absolute and ordained power is crucial for Gassendi, I know only a few passages where he actually draws the distinction.

This is not an insurmountable problem. The absence of much explicit statement of a view may indicate that it is a fundamental assumption too obvious to need stating. However, Osler holds that one of the chief points of difference between Gassendi and Descartes is Gassendi's acceptance of voluntarism and Descartes's rejection of it, and she takes Gassendi's opposition to Cartesian a priorism to stem in large part from his voluntarism.[29] If this is the main source of such opposition, however, it is rather odd that the claim is never articulated.

Moreover, even if we grant Osler that voluntarism and a certain style of science are very closely connected, we must still ask whether Gassendi adopted his style of science because of his voluntarism or vice versa.[30] One might think manuscript evidence could answer this question. Bloch holds that theological material like the account of the relation between God's will and his intellect is inserted after 1641. In opposition to this, Osler appeals to the important 1631 letter to Peiresc where Gassendi outlines the Epicurus project as it then stood. This outline adumbrates sections on the existence and nature of God, providence, and the immortality of the soul.[31] She takes this letter to "provide strong evidence against Bloch's contention that Gassendi interpolated the theological material into the

[29] It is standard to read Descartes as a voluntarist, given his views on the creation of the eternal truths. However, Osler, *Divine Will and the Mechanical Philosophy*, 146–52, objects to this characterization and argues that because the Cartesian God restricts his freedom of action *after* creation by creating necessary laws of nature, he should be understood as an intellectualist (148). Her definitions of voluntarism and intellectualism rely heavily on the distinction between the absolute and ordinary divine powers (17–18).

[30] I owe this point to Wilson, "Theological Foundations for Modern Science?" 601.

[31] Peiresc, *Lettres de Peiresc*, 4.249 ff.

Physics in the early 1640s."[32] But aside from being entirely consistent with Bloch's claim that the pre-1641 material simply juxtaposes Epicurean and orthodox themes with no integration, it also is not sufficient to show that Gassendi was a voluntarist at this point, let alone that voluntarism motivated the development of his natural philosophy over the next twenty years.

Earlier I mentioned Bloch's point that despite his good theological eductation, Gassendi does not seem to have much interest in theology. Bloch takes the absence of a developed theology as evidence for his incompatibility thesis: If Gassendi indeed cannot reconcile faith and reason, he would be on very shaky ground developing a theology. Osler, on the other hand, insists that theology is present underneath the surface even where it is not made explicit. Both Bloch and Osler assume that if Gassendi's Christian beliefs are integral to his philosophical system, then they must be reflected in theological engagement. We should be suspicious of this assumption. It implies that *any* writer in the period who does not discuss theology explicitly has strong subterranean commitments to an entirely implicit theology, to the doctrine of double truth, or to atheism. The assumption thus leaves no room for a division of intellectual labor. But it is by no means obvious that every philosopher committed to religious orthodoxy should thereby be committed to putting forth a detailed theology. Indeed, if we take Gassendi's overarching project to be the dissemination of an atomist program, it makes sense for him to avoid tying that program to any disputed theological claims. Gassendi's reluctance to become involved in theological debate thus accords well with his epistemic pessimism and his main philosophical goals.

[32] Osler, *Divine Will and the Mechanical Philosophy*, 27.

Bibliography

I. Gassendi's Works

I have cited only those manuscripts, published works, and translations that I have actually consulted. Readers in search of a complete bibliography of Gassendi's works should consult Centre International d'Études Gassendiennes, *Catalogue: Pierre Gassendi.* References not otherwise identified are to Gassendi's *Opera Omnia,* cited by volume, number, and (where applicable) column. Readers wishing to know the work being cited should consult "References to Gassendi's Works" at the beginning of this book.

British Library MS Harley 1677 [portions of the work known as *De vita et doctrina Epicuri*], 1634. British Library, London.

De vita et moribus Epicuri libri octo. Lyon: Guillaume Barbier, 1647.

Animadversiones in decimum librum Diogenis Laertii, qui est de vita, moribus, placitisque Epicuri. Lyon: Guillaume Barbier, 1649.

Receuil de lettres des sieurs Morin, de la Roche, de Neure et Gassend: en suite l'apologie du Sieur Gassend touchant la question De motu impresso a motore translato. Paris: Augustin Courbé, 1650.

Institutio astronomica, juxta hypotheseis tam veterum, quam Copernici et Tychonis, dictata a Petro Gassendo regio matheseos professore. London: Jacob Flesher, 1653.

The Mirrour of True Nobility & Gentility. Being the life of the renowned Nicolaus Claudius Fabricius Lord of Peiresk, Senator of the Parliament at Aix. Written by the learned Petrus Gassendus, professor of the mathematicks to the King of France. Englished by W. Rand, Doctor of Physick. London: J. Streater for Humphrey Moseley, 1657.

Petri Gassendi Opera Omnia in sex tomos divisa. 6 volumes. Lyon: Laurent Anisson and Jean-Baptiste Devenet, 1658. Reprinted in facsimile with an introduction by Tullio Gregory. Stuttgart-Bad Canstatt: Friedrich Frohmann, 1964.

The Vanity of Judiciary Astrology, or, Divination by the stars. Lately written in Latin, by that great schollar and mathematician, the illustrious Petrus Gassendus, mathematical professor to the king of France. Translated into English by a person of quality. London: Humphrey Moseley, 1659.

Institutio logica, et Philosophiae Epicuri Syntagma. Authore V. Cl. Petro Gassendi. London: John Redmayne, 1668.
A Discourse on the Antiquity, Progress, and Augmentation of Astronomy. London: William Hawes, 1699.
Lettres familières à François Luillier pendant l'hiver 1632–1633. Edited by Bernard Rochot. Paris: J. Vrin, 1944.
Dissertations en forme de paradoxes contre les Aristotéliciens (Exercitationes paradoxicae adversus Aristoteleos) Livres I et II. Edited and translated by Bernard Rochot. Paris: J. Vrin, 1959.
Disquisitio metaphysica; seu, Dubitationes et instantiae adversus Renati Cartesii Metaphysicam et responsa. Recherches métaphysiques; ou, Doutes et instances contre la Métaphysique de R. Descartes et ses réponses. Edited and translated by Bernard Rochot. Paris: J. Vrin, 1962.
The Selected Works of Pierre Gassendi. Edited and translated by Craig G. Brush. New York: Johnson Reprint, 1972.
Pierre Gassendi's Institutio Logica (1658). Edited and translated by Howard Jones. Assen: Van Gorcum, 1981.
Vie et Moeurs d'Epicure. 2 volumes. Translated by Sylvie Taussig. Paris: Editions Alive, 2001.
Pierre Gassendi (1592–1655): Lettres latines. 2 volumes. Translated by Sylvie Taussig. Turnhout: Brepols, 2004.

II. Other Primary Sources

Aquinas, Thomas. *The Summa Contra Gentiles of Saint Thomas Aquinas.* London: Burns Oates and Washbourne, 1923.
Summa theologiae. London: Blackfriars, 1964.
A Commentary on Aristotle's De Anima. Translated by Robert Pasnau. New Haven: Yale University Press, 1999.
Aristotle. *The Complete Works of Aristotle.* 2 volumes. Edited by Jonathan Barnes. Princeton: Princeton University Press, 1984.
Arnauld, Antoine. *On True and False Ideas.* Translated by Elmar J. Kremer. Lewiston, NY: Lampeter, 1990.
Arnauld, Antoine, and Pierre Nicole. *Logic, or, The Art of Thinking.* Translated by Jill Vance Buroker. Cambridge: Cambridge University Press, 1996.
Augustine. *Against the Academicians; The Teacher.* [*Contra Academicos.*] Translated by Peter King. Indianapolis: Hackett, 1995.
Bacon, Francis. *The Advancement of Learning.* Edited by Arthur Johnston. Oxford: Clarendon Press, 1974.
Baillet, Adrien. *La vie de Monsieur Des-Cartes.* 2 volumes. Paris: Daniel Horthemels, 1964.
Basso, Sebastien. *Philosophia naturalis adversus Aristotelem, in quibus abstrusa veterum Physiologia restauratur, & Aristotelis errores solidis rationibus refelluntur.* Geneva: Pierre de la Rouiere, 1621.
Bayle, Pierre. *Dictionnaire historique et critique.* 3rd edition. Rotterdam: Michel Böhm, 1720.

Bayle, Pierre. *Historical and Critical Dictionary.* Translated by Richard Popkin. Indianapolis: Hackett, 1991.

Beeckman, Isaac. *Journal tenu par Isaac Beeckman de 1604–1634.* Edited by Cornelius de Waard. The Hague: Martinus Nijhoff, 1939–53.

Berkeley, George. *Works of George Berkeley.* Edited by A. A. Luce and T. E. Jessop. 9 volumes. London: Nelson, 1948–57.

Bernier, François. *Abrégé de la philosophie de M. Gassendi.* 8 volumes. Lyon: Anisson, Posuel & Rigaud, 1678. Reprint edition, edited by Sylvia Murr and Geneviève Stefani. Paris: Fayard, 1992.

Three Discourses of Happiness, Virtue, and Liberty, collected from the works of the Learn'd Gassendi. London: Awnshawm & John Churchill, 1699.

Bougerel, Joseph. *Vie de Pierre Gassendi, prévôt de l'église de Digne & professeur de mathématiques au Collège Royal.* Paris: Jacques Vincent, 1737. Reprint, Geneva: Slatkine Reprints, 1970.

Burgersdijck, Franco. *Idea philosophiae tum moralis, tum naturalis. Sive, Epitome compendiosa utriusque ex Aristotele excerpta, & methodice disposita.* Oxford: Joseph Godwin and Richard Davis, 1667.

Franconis Burgersdici Institionum metaphysicarum lib. ii. London: J. Creek and J. Baker, 1653.

Campanella, Tommaso. *De Sensu Rerum et Magia, Libri Quatuor.* Frankfurt: Godefrid Tampachius, 1620.

A Defence of Galileo the Mathematician from Florence. Edited and translated by Richard J. Blackwell. South Bend, IN: University of Notre Dame Press, 1994.

Charleton, Walter. *The Darkness of Atheism dispelled by the Light of Nature: A physico-theologicall treatise.* London: J.F. for William Lee, 1652.

Physiologia Epicuro–Gassendo–Charletoniana: or, A fabrick of science natural, upon the hypothesis of atoms, founded by Epicurus, repaired by Petrus Gassendus, augmented by Walter Charleton, Dr. in medicine, and physician to the late Charles, Monarch of Great-Britain. London: Tho. Newcomb for Thomas Heath, 1654.

Epicurus's morals, collected partly out of his owne Greek text, in Diogenes Laertius, and partly out of the rhapsodies of Marcus Antoninus, Plutarch, Cicero, & Seneca. And faithfully Englished. London: William Wilson for Henry Herringman, 1656.

The immortality of the human soul, demonstrated by the light of nature. In two dialogues. London: William Wilson for Henry Herringman, 1657.

Cicero, Marcus Tullius. *De natura deorum. Academica.* Translated by H. Rackham. Loeb Classical Library. Cambridge, MA: Harvard University Press, 1940.

Clave, Étienne de. *Nouvelle lumière philosophique des vrais principes et élémens de nature, & qualité d'iceux.* Edited by Bernard Joly. Paris: Fayard, 2000.

Le cours de chimie, second liure des principes de nature. Paris: Olivier de Varennes, 1646.

Colegio das Artes (Coimbra, Portugal). *Commentarii Colegii Conimbricensis Societatis Iesu, in octo libros Physicorum Aristotelis Stagiritae.* London: Horatio Cardon, 1602.

Commentarii Collegii Conimbricensis Societatis Iesu, in tres libros De anima Aristotelis Stagiritae. Cologne: Lazarus Zetner, 1617.

Cudworth, Ralph. *The True Intellectual System of the Universe: The first part, wherein, all the reason and philosophy of atheism is confuted; and its impossibility demonstrated.* London: Richard Royston, 1678.

Daniel, Gabriel. *Voiage du monde de Descartes.* Paris: Veuve de S. Benard, 1691.
A voyage to the world of Cartesius. London: Thomas Bennet, 1692.

Descartes, René. *The Philosophical Writings of Descartes.* 3 volumes. Translated by John Cottingham, Robert Stoothoff, Dugald Murdoch, and (3rd volume only) Anthony Kenny. Cambridge: Cambridge University Press, 1984–91.
Oeuvres de Descartes. 11 volumes. Edited by Charles Adam and Paul Tannery. Paris: J. Vrin, 1996.

Diogenes Laertius. *Lives of Eminent Philosophers.* Edited by R. D. Hicks. Loeb Classical Library. London: W. Heinemann, 1925.

Digby, Kenelm. *Two treatises, in the one of which, the Nature of Bodies, in the other, the Nature of Mans Soule, is looked into, in way of discovery, of the immortality of reasonable soules.* Paris: Gilles Blaizot, 1644.

Du Moulin, Pierre. *La philosophie, mise en français, et divisée entre trois parties, sçavoir, elements de la logique, la physique ou science naturalle, l'ethyque ou science morale.* Paris: Thomas Blaise and Olivier de Varenne, 1644.

Dupleix, Scipion. *La Métaphysique.* Edited by Roger Ariew. Paris: Fayard, 1992.
La Physique. Edited by Roger Ariew. Paris: Fayard, 1992.

Epicurus. *Epicurus, the Extant Remains.* Edited and translated by Cyril Bailey. Hildesheim: G. Olms, 1974.
The Epicurus Reader: Selected Testimonia and Writings. Translated and edited by Brad Inwood and Lloyd Gerson. Indianapolis: Hackett, 1994.

Eustachius a Sancto Paolo. *Summa philosophiae quadripartita: de rebus dialecticis, moralibus, physicis, & metaphysicis.* Paris: Pierre Bilaine, 1620.

Fabricius, Hieronymus. *The embryological treatises of Hieronymus Fabricius of Aquapendente: The formation of the egg and of the chick (De formatione ovi et pulli), The formed fetus (De formato fetus).* Edited and translated by Howard B. Adelmann. Ithaca, NY: Cornell University Press, 1942.

Fernel, Jean. *The Physiologia of Jean Fernel (1567).* Edited and translated by J. M. Forrester. Philadelphia: American Philosophical Society, 2003.

Ficino, Marsilio. *Platonic Theology.* Edited and translated by Michael J. B. Allen with John Warden. Cambridge, MA: Harvard University Press, 2001.

Fleury, Marie-Antoinette, and Georges Bailhache (editors). "Documents inédits sur Gassendi." In *Pierre Gassendi, 1592–1655.* Edited by Bernard Rochot. Paris: Editions Albin Michel, 1955.

Fludd, Robert. *Utriusque cosmi maioris scilicet et minoris metaphysica, physica atque technica historia.* Oppenheim: Johan-Theodor de Bry, 1617.
Sophia cum moria certamen, in quo lapis Lydius a falso structore, fr. Marino Mersenno reprobatus, celeberrima voluminis sui Babylonici, in Genesim, figmenta accurate examinat. Frankfurt: Joachim Frizius, 1629.
Doctor Fludds answer unto m. Foster, or, The squeesing of parson Fosters sponge, ordained by him for the wiping away of the weapon-salve. London: N. Butter, 1631.
Clavis philosophiae et alchymiae Fluddanae, sive, Roberti Fluddi. Ad epistolicam Petri Gassendi Theologi exercitationem responsum. Frankfurt: William Fitzerum, 1633.
Philosophia Moysaica. Gouda: Peter Rammazenius, 1638.
Mosaicall philosophy: grounded upon the essential truth or eternal sapience. London: Humphrey Moseley, 1659.
Robert Fludd and his Philosophicall Key. Edited by Allen Debus. New York: Science History Publications, 1979.

Foster, William. *Hoplocrisma-spongus, or, A sponge to wipe away the weapon-salve.* London: Thomas Cotes for John Grove, 1631.

Galen. *Selected Works.* Edited and translated by P. N. Singer. Oxford: Oxford University Press, 1977.

On the Doctrines of Hippocrates and Plato. Edited and translated by Phillip de Lacy. 2 volumes. Berlin: Akademie-Verlag, 1980.

Galilei, Galileo. *Discoveries and Opinions of Galileo.* Translated by Stillman Drake. Garden City, NY: Doubleday Anchor, 1957.

Dialogue Concerning the Two Chief World Systems. Berkeley: University of California Press, 1967.

Two New Sciences: Including Centers of Gravity and Force of Percussion. Translated by Stillman Drake. Madison: University of Wisconsin Press, 1974.

Gilbert, William. *De Magnete, Magnetisque Corporibus, et de magno magnete tellure.* London: Peter Short, 1600.

Goclenius, Rudolf. *Lexicon philosophicum: quo tanquam clave philosophiae fores aperiuntur.* Frankfurt: 1613. Reprinted Hildesheim: Georg Olms, 1964.

Harvey, William. *Anatomical Exercitations, concerning the generation of living creatures.* London: 1653.

Disputations Touching the Generation of Animals. Oxford: Blackwell Scientific, 1981.

Hill, Nicholas. *Philosophia Epicurea, Democriteana, Theophrastica proposita simpliciter, non edocta.* Paris: R. Thierry, 1601.

Hobbes, Thomas. *Elementorum Philosophiae Sectio Prima De Corpore.* London: Andreas Crook, 1655.

Opera philosophica quae latine scripsit. 5 volumes. Edited by William Molesworth. London: Bohn, 1839–45.

Thomas White's De Mundo Examined. London: Bradford University Press, 1976.

The Correspondence. 2 volumes. Edited by Noel Malcolm. Oxford: Clarendon Press, 1994.

Horace. *Satires, Epistles and Ars Poetica.* Translated by H. Rushton Fairclough. Loeb Classical Library. Cambridge, MA: Harvard University Press, 1926.

Hume, David. *Enquiries concerning Human Understanding and concerning the Principles of Morals.* Edited by L. A. Selby-Bigge, 3rd edition revised by P.H. Nidditch. Oxford: Clarendon Press, 1975.

A Treatise of Human Nature. Edited by L. A. Selby-Bigge, 2nd edition revised by P. H. Nidditch. Oxford: Oxford University Press, 1978.

Huygens, Christian. *Oeuvres complètes par la Societé Hollandaise des sciences.* 22 volumes. The Hague: Martinus Nijhoff, 1888–1950.

Keckermann, Bartholomew. *Systema Compendiosum Totius Mathematices, hoc est Geometriae, Optice, Astronomiae, et Geographiae.* Oxford: William Hall for Francis Oxlad, 1651.

La Poterie, Antoine de. "Memoires Touchant la Naissance, Vie et Moeurs de Gassendi." Edited by Philippe Tamizey de Larroque. *Revue des Questions Historique* 22 (1877), 211–40.

Leibniz, G. W. *Philosophical Papers and Letters.* Edited and translated by Leroy Loemker. Chicago: University of Chicago Press, 1956.

Die philosophischen Schriften von Gottfried Wilhelm Leibniz. 7 volumes. Edited by C. I. Gerhardt. Hildesheim: Georg Olms, 1960–1.

Lipsius, Justus. *J. L. Manuductionis ad Stoicam philosophiam libri tres.* Paris: Hadrian Perier, 1604.

J. Lipsii Physiologiae Stoicorum libri tres. Paris: Hadrian Perier, 1604.

Locke, John. *An Essay Concerning Human Understanding.* Edited by Peter H. Nidditch. Oxford: Oxford University Press, 1975.

Drafts for the 'Essay Concerning Human Understanding' and Other Philosophical Writings. Edited by P. H. Nidditch and G. A. J. Rogers. Oxford: Clarendon Press, 1990.

Lucretius Carus, Titus. *De rerum natura.* Edited by Cyril Bailey. Oxford: Clarendon Press, 1947.

Malebranche, Nicolas. *Oeuvres complètes.* 20 volumes. Edited by André Robinet. Paris: J. Vrin, 1958–84.

Marx, Karl, and Friedrich Engels. *The Holy Family.* Translated by R. Divson. Moscow: Foreign Languages Publishing House, 1956.

Mersenne, Marin. *La vérité des sciences contre les sceptiques ou pyrrhoniens.* Paris: 1625. Facsimile reprint, Stuttgart-Bad Cannstatt: Friedrich Frommann, 1969.

Correspondance du P. Marin Mersenne. 15 volumes. Edited by Cornelius de Waard and Armand Beaulieu. Paris: CNRS, 1933–83.

Molina, Luis de. *Liberi Arbitrii cum Gratiae Donis, Divina Praescientia, Providentia, Praedestinatione et Reprobatione Concordia.* Edited by J. Rabeneck. Ona: Collegium maximum Societatis Iesu, 1953.

More, Henry. *Henry More's Manual of Metaphysics: A translation of the Enchiridion Metaphysicum (1679).* Translated by Alexander Jacob. Hildesheim: Georg Olms, 1995.

Moxon, Joseph. *A Tutor to Astronomy and Geography, Or, an Easie and Speedy Way to Know the Use of Both of the Globes, Coelestial and Terrestrial.* London: Phillip Lea, 1699.

Newton, Isaac. *Unpublished Scientific Papers of Isaac Newton.* Edited by Rupert Hall and Marie Boas Hall. Cambridge: Cambridge University Press, 1962.

Certain Philosophical Questions: Newton's Trinity Notebook. Edited by J. E. McGuire and Martin Tamny. Cambridge: Cambridge University Press, 1983.

Philosophical Writings. Edited by Andrew Janiak. Cambridge: Cambridge University Press, 2004.

Pascal, Blaise. *The Physical Treatises of Pascal: The Equilibrium of Liquids and the Weight of the Mass of the Air.* Translated by A. G. H. Spiers. New York: Columbia University Press, 1937.

Oeuvres complètes. 4 volumes. Edited by Jean Mesnard. Paris: Desclée de Brouwer, 1964–92.

Patrizi, Francesco. *Discussionum peripateticarum tomi IV, quibus Aristotelicae philosophiae universa historia atque dogmata nunc veterum placitis collata, eleganter et erudite declarantur.* Basel: ad Pernam Lecythum, 1581.

Nova de Universis Philosophia. Ferrara: Benedict Mammarelli, 1591.

Peiresc, Nicolas-Claude Fabri. *Lettres de Peiresc.* 7 volumes. Edited by Philippe Tamizey de Larroque. Paris: Imprimerie Nationale, 1893.

Pemble, William. *De Origine Formarum.* Cambridge: Roger Daniel for John Bartlett, 1650.

Philodemus. *Philodemus: On Methods of Inference.* Edited and translated by Phillip de Lacey and Estelle de Lacey. Naples: Bibliopolis, 1978.

Raphson, Joseph. *De Spatio Reali seu Ente Infinito.* London: John Taylor, 1697.
Sanderson, Robert. *Logicae Artis Compendium.* 2nd edition. Oxford: John Lichfield and Jacob Short, 1618.
Sennert, Daniel. *Epitome Naturalis Scientiae.* Oxford: John Lichfield and Henry Cripps, 1632.
Hypomnemata Physica. Frankfurt: Clement Schleich, 1636.
Severinus, Petrus. *Idea Medicinae Philosophicae.* Basel: Henricpetrus, 1571.
Sextus Empiricus. *Adversus Mathematicos.* Loeb Classical Library. 4 volumes. Edited and translated by R.G. Bury. Cambridge, MA: Harvard University Press, 1968.
Outlines of Scepticism. Translated by Julia Annas and Jonathan Barnes. Cambridge: Cambridge University Press, 2000.
Stanley, Thomas. *The History of Philosophy: Containing the Lives, Opinions, Actions and Discourses of the Philosophers of Every Sect.* 3 volumes. London: Thomas Bassett, 1687.
Suárez, Francisco. *Disputationes metaphysicae.* 2 volumes. Reprint Hildesheim: Georg Olms, 1965.
On Efficient Causality: Metaphysical Disputations 17, 18 and 19. Translated by A. J. Freddosso. New Haven: Yale University Press, 1994.
On the Formal Cause of Substance: Metaphysical Disputation XV. Translated by John Kronen. Milwaukee: Marquette University Press, 2000.
On Creation, Conservation, and Concurrence: Metaphysical Disputations 20, 21, and 22. Translated by A. J. Freddosso. South Bend, IN: St. Augustine's Press, 2002.
Telesio, Bernardino. *De rerum natura iuxta propria principia.* Naples: I. Cacchius, 1570.
Toletus, Francisco. *Commentaria una cum Quaestionibus in Octo Libros Aristotelis de Physica auscultatione. Item in Lib. Arist. De Generatione et Corruptione.* Rome: I. Martinelli, 1590.
Van Helmont, Jean-Baptiste. *Ortus medicinae. Id est, initia physicae inaudita. Progressus medicinae novus, in morborum ultionem, ad vitam longam.* Amsterdam: Ludovic Elzevier, 1648.

III. Secondary Sources

Adelmann, Howard B. *Marcello Malpighi and the Evolution of Embryology.* Ithaca, NY: Cornell University Press, 1966.
Alexander, Peter. *Ideas, Qualities and Corpuscles.* Cambridge: Cambridge University Press, 1985.
Algra, Keimpe, Jonathan Barnes, Jaap Mansfeld, and Malcolm Schofield (editors). *The Cambridge History of Hellenistic Philosophy.* Cambridge: Cambridge University Press, 1999.
Allen, James. *Inference from Signs: Ancient Debates about the Nature of Evidence.* Oxford: Oxford University Press, 2001.
Annas, Julia. "Epicurus' Philosophy of Mind." In *Companions to Ancient Thought II: Psychology.* Edited by Stephen Everson. Cambridge: Cambridge University Press, 1991.
Hellenistic Philosophy of Mind. Berkeley: University of California Press, 1992.
Annas, Julia, and Jonathan Barnes. *The Modes of Scepticism: Ancient Texts and Modern Interpretations.* Cambridge: Cambridge University Press, 1985.

Ariew, Roger. "Damned If You Do: Cartesians and Censorship, 1663–1706." *Perspectives on Science* (1994), 255–74.

Descartes and the Last Scholastics. Ithaca, NY: Cornell University Press, 1999.

Ariew, Roger, and Marjorie Grene. *Descartes and His Contemporaries: Meditations, Objections and Replies.* Chicago: University of Chicago Press, 1995.

Armogathe, Jean-Robert. "L'Enseignement de Pierre Gassendi au Collège Royal d'Aix-en-Provence et la Traditions Philosophique des Grands Carmes." In *Gassendi et l'Europe, 1592–1792: actes du colloque international de Paris.* Edited by Sylvia Murr. Paris: J. Vrin, 1997.

"Proofs of the Existence of God." In *The Cambridge History of Seventeenth-Century Philosophy.* Edited by Daniel Garber and Michael Ayers. Cambridge: Cambridge University Press, 1998.

Ashworth, E. J. *Language and Logic in the Post-Medieval Period.* Dordrecht: Reidel, 1974.

Studies in Post-Medieval Semantics. London: Variorum Reprints, 1985.

"Traditional Logic." In *The Cambridge History of Renaissance Philosophy.* Edited by C. B. Schmitt and Quentin Skinner. Cambridge: Cambridge University Press, 1988.

Asmis, Elizabeth. *Epicurus' Scientific Method.* Ithaca, NY: Cornell University Press, 1984.

Ayers, Michael. *Locke: Epistemology and Ontology.* New York: Routledge, 1991.

"Ideas and Objective Being." In *The Cambridge History of Seventeenth-Century Philosophy.* Edited by Daniel Garber and Michael Ayers. Cambridge: Cambridge University Press, 1998.

"Theories of Knowledge and Belief." In *The Cambridge History of Seventeenth-Century Philosophy.* Edited by Daniel Garber and Michael Ayers. Cambridge: Cambridge University Press, 1998.

Barber, Kenneth, and Jorge Gracia. *Individuation and Identity in Early Modern Philosophy: Descartes to Kant.* Albany: State University of New York Press, 1994.

Baumgartner, F. J. "Galileo's French Correspondents." *Annals of Science* 45 (1988): 169–82.

Beaulieu, Armand. "Les Réactions des Savants Français au Debut du XVII Siècle devant l'Héliocentrisme de Galilée." In *Novità celesti e crisi del sapere.* Edited by Paolo Galluzzi. Florence: Giunta Barbera, 1984.

Berr, Henri. *Du scepticisme de Gassendi.* Translated by Bernard Rochot. Paris: Albin Michel, 1960.

Blackwell, Constance, and S. Kusukawa (editors). *Philosophy in the Sixteenth and Seventeenth Centuries: Conversations with Aristotle.* Brookfield, VT: Ashgate, 1999.

Blackwell, Richard J. *Galileo, Bellarmine, and the Bible.* South Bend, IN: University of Notre Dame Press, 1991.

Bloch, Olivier. *La philosophie de Gassendi: nominalisme, matérialisme, et métaphysique.* The Hague: Martinus Nijhoff, 1971.

"Gassendi and the Transition from the Middle Ages to the Classical Era." *Yale French Studies* 49 (1973), 43–55.

Brandt, Reinhard. "Historical Observations on the Genesis of the Three-Dimensional Optical Picture (Gassendi, Locke, Berkeley)". *Ratio* 17.1 (1975), 176–90.

Brett, G. S. *The Philosophy of Gassendi*. London: Macmillan, 1908.

Brockliss, L. W. B. *French Higher Education in the Seventeenth and Eighteenth Centuries: A Cultural History*. Oxford: Oxford University Press, 1987.

"Descartes, Gassendi, and the Reception of the Mechanical Philosophy in the French Collèges de Plein Exercise, 1640–1730." *Perspectives on Science* 3.4 (1995), 450–79.

Brundell, Barry. *Pierre Gassendi: From Aristotelianism to a New Natural Philosophy*. Dordrecht: Reidel, 1987.

Burnyeat, Miles. "Idealism and Greek Philosophy: What Descartes Saw and Berkeley Missed." *The Philosophical Review* 91 (1982), 3–40.

Centre International d'Études Gassendiennes. *Catalogue: Pierre Gassendi*. Digne-les-Bains: Centre International d'Études Gassendiennes, 1992.

Chappell, Vere. *Essays on Early Modern Philosophers: Grotius to Gassendi*. New York: Garland, 1992.

Clark, J. T. "Pierre Gassendi and the Physics of Galileo." *Isis* 54 (1963), 351–70.

Clarke, Desmond. *Occult Powers and Hypotheses: Cartesian Natural Philosophy under Louis XIV*. Oxford: Clarendon Press, 1989.

Clatterbaugh, Kenneth. *The Causation Debate in Early Modern Philosophy, 1637–1739*. New York: Routledge, 1999.

Clericuzio, Antonio. *Elements, Principles and Corpuscles: A Study of Atomism and Chemistry in the Seventeenth Century*. Dordrecht: Kluwer, 2000.

"Gassendi, Charleton and Boyle on Matter and Motion." In *Late Medieval and Early Modern Corpuscular Matter Theories*. Edited by Christoph Lüthy, John Murdoch, and William Newman. Leiden: Brill, 2001.

Clucas, Stephen. "The Atomism of the Cavendish Circle: A Reappraisal." *The Seventeenth Century* 9.2 (1994), 247–68.

"The Infinite Variety of Formes and Magnitudes: 16th and 17th Century English Corpuscular Philosophy and Aristotelian Theories of Matter and Form." *Early Science and Medicine* 2.3 (1997), 257–71.

"Corpuscular Matter Theory in the Northumberland Circle." In *Late Medieval and Early Modern Corpuscular Matter Theories*. Edited by Christoph Lüthy, John Murdoch, and William Newman. Leiden: Brill, 2001.

Crocker, Robert. *Henry More, 1614–1687: A Biography of the Cambridge Platonist*. Dordrecht: Kluwer, 2003.

Cunningham, Andrew. "The Identity of Natural Philosophy: A Response to Edward Grant." *Early Science and Medicine* 5.3 (2000), 259–78.

Darmon, Jean-Claude. "Gassendi et la 'rhetorique' de Descartes." *Papers on French Seventeenth Century Literature* 25.49 (1998), 401–29.

Philosophie épicurienne et littérature au XVIIe siècle en France. Paris: Presses Universitaires de France, 1998.

Dear, Peter. *Mersenne and the Learning of the Schools*. Ithaca, NY: Cornell University Press, 1988.

Discipline and Experience. Chicago: University of Chicago Press, 1995.

"Method and the Study of Nature." In *The Cambridge History of Seventeenth-Century Philosophy*. Edited by Daniel Garber and Michael Ayers. Cambridge: Cambridge University Press, 1998.

Debus, Allen. *The French Paracelsans: The Chemical Challenge to Medical and Scientific Tradition in Early Modern France.* Cambridge: Cambridge University Press, 1991.

Des Chene, Dennis. *Physiologia: Natural Philosophy in Late Aristotelian and Cartesian Thought.* Ithaca, NY: Cornell University Press, 2000.

Life's Forms. Ithaca, NY: Cornell University Press, 2001.

Spirits and Clocks: Machine and Organism in Descartes. Ithaca, NY: Cornell University Press, 2001.

Duchesneau, François. *Les modèles du vivant de Descartes à Leibniz.* Paris: J. Vrin, 1998.

Dutton, Blake. "Physics and Metaphysics in Descartes and Galileo." *Journal of the History of Philosophy* 37.1 (1999), 49–71.

Easton, Patricia (editor). *Logic and the Workings of the Mind: The Logic of Ideas and Faculty Psychology in Early Modern Philosophy.* Atascadero, CA: Ridgeview, 1997.

Egan, Howard. *Gassendi's View of Knowledge: A Study of the Epistemological Basis of His Knowledge.* Lanham, MD: University Press of America, 1984.

Emerton, Norma E. *The Scientific Reinterpretation of Form.* Ithaca, NY: Cornell University Press, 1984.

Everson, Stephen. "Epicurus on the Truth of the Senses." In *Companions to Ancient Thought I: Epistemology.* Edited by Stephen Everson. Cambridge: Cambridge University Press, 1990.

"Epicurus on Mind and Language." In *Companions to Ancient Thought III: Language.* Edited by Stephen Everson. Cambridge: Cambridge University Press, 1994.

Findlen, Paula (editor). *Athanasius Kircher: The Last Man who Knew Everything.* New York: Routledge, 2004.

Fine, Gail. "Descartes and Ancient Skepticism: Reheated Cabbage?" *The Philosophical Review* 109.2 (2000), 195–234.

Fisher, Saul. "Science and Skepticism in the 17th Century: The Atomism and Scientific Method of Pierre Gassendi." PhD dissertation, CUNY, 1997.

"Gassendi's Atomist Account of Generation and Heredity in Plants and Animals." *Perspectives on Science* 11.4 (2003), 484–512.

Floridi, Luciano. *Sextus Empiricus: The Transmission and Recovery of Pyrrhonism.* Oxford: Oxford University Press, 2002.

Fouke, Daniel C. "'Mechanical' and 'Organical' Models in 17th Century Models of Biological Reproduction." *Science in Context* 3.2 (1989), 366–81.

"Pascal's Physics." In *The Cambridge Companion to Pascal.* Edited by Nicholas Hammond. Cambridge: Cambridge University Press, 2003.

Fowler, D., and P. G. Fowler. *Lucretius on Atomic Motion: A Commentary on De Rerum Natura.* Oxford: Oxford University Press, 2002.

Frank, Robert G. *Harvey and the Oxford Physiologists.* Berkeley: University of California Press, 1980.

Freddoso, Alfred J. "God's General Concurrence with Secondary Causes: Why Conservation Is Not Enough." *Philosophical Perspectives* 5 (1991), 553–85.

"God's General Concurrence with Secondary Causes: Pitfalls and Prospects." *American Catholic Philosophical Quarterly* 67 (1994), 131–56.

French, R. K. *William Harvey's Natural Philosophy.* Cambridge: Cambridge University Press, 1994.

Freudenthal, Gad. "Stoic Concepts in Mechanical Philosophy: The Problem of Electrical Attraction." In *Renaissance and Revolution: Humanists, Scholars, Craftsmen and Natural Philosophers in Early Modern Europe.* Edited by J. V. Field and James Frank. Cambridge: Cambridge University Press, 1993.

Fuchs, T., and M. G. Grene. *The Mechanization of the Heart: Harvey and Descartes.* Rochester, NY: University of Rochester Press, 2001.

Furley, David J. "Aristotle and the Atomists on Motion in a Void." In *Motion and Time, Space and Matter.* Edited by Peter K. Machamer and R. G. Turnbull. Columbus: Ohio State University Press, 1976.

Gabbey, Alan. "New Doctrines of Motion." In *The Cambridge History of Seventeenth-Century Philosophy.* Edited by Daniel Garber and Michael Ayers. Cambridge: Cambridge University Press, 1998.

Gabbey, Alan, and Roger Ariew. "Body: The Scholastic Background." In *The Cambridge History of Seventeenth-Century Philosophy.* Edited by Daniel Garber and Michael Ayers. Cambridge: Cambridge University Press, 1998.

Galluzzi, Paolo. "Gassendi and l'Affaire Galilée of the Laws of Motion." *Science in Context* 13.3–4 (2000), 509–45.

Garber, Daniel. "Semel in Vita." In *Essays on Descartes' Meditations.* Edited by Amelie Rorty. Berkeley: University of California Press, 1986.

Descartes' Metaphysical Physics. Chicago: University of Chicago Press, 1992.

"Descartes and Occasionalism." In *Causation in Early Modern Philosophy.* Edited by Steven Nadler. University Park: Pennsylvania State University Press, 1993.

Gatti, Hillary. "Giordano Bruno's Soul-Powered Atoms." In *Late Medieval and Early Modern Corpuscular Matter Theories.* Edited by Christoph Lüthy, John Murdoch, and William Newman. Leiden: Brill, 2001.

Gaukroger, Stephen. *Descartes: An Intellectual Biography.* Oxford: Clarendon Press, 1995.

Francis Bacon and the Transformation of Early Modern Philosophy. Cambridge: Cambridge University Press, 2001.

Gaukroger, Stephen. *Descartes' System of Natural Philosophy.* Cambridge: Cambridge University Press, 2002.

Glidden, David. "Hellenistic Background for Gassendi's Theory of Ideas." *Journal of the History of Ideas* 49 (1988), 405–24.

"Parrots, Pyrrhonists and Native Speakers." In *Companions to Ancient Thought III: Language.* Edited by Stephen Everson. Cambridge: Cambridge University Press, 1994.

Gorman, M. J. "A Matter of Faith? Christoph Scheiner, Jesuit Censorship, and the Trial of Galileo." *Perspectives in Science* 4.3 (1996), 283–320.

Grant, Edward. "The Condemnation of 1277, God's Absolute Power, and Physical Thought in the Late Middle Ages." *Viator: Medieval and Renaissance Studies* 10 (1979), 211–44.

Much Ado About Nothing: Theories of Space and the Vacuum from the Middle Ages to the Scientific Revolution. Cambridge: Cambridge University Press, 1981.

"The Condemnation of 1277." In *The Cambridge History of Later Medieval Philosophy.* Edited by Norman Kretzman, Anthony Kenny, and Jan Pinborg. Cambridge: Cambridge University Press, 1982.

"God and Natural Philosophy: The Late Middle Ages and Sir Isaac Newton." *Early Science and Medicine* 5.3 (2000), 279–300.

Gregory, Tullio. "Libertinisme Érudit in Seventeenth-Century France and Italy: The Critique of Ethics and Religion." *British Journal for the History of Philosophy* 6.3 (1998), 323–50.

Grene, Marjorie. *Descartes.* Minneapolis: University of Minnesota Press, 1985.

Guerlac, Henri. "Can There Be Colors in the Dark?" *Journal of the History of Ideas* 47.1 (1986), 3–20.

Hall, A. Rupert. *Henry More: Magic, Religion and Experiment.* Cambridge, MA: Blackwell, 1990.

Hatfield, Gary. "Force (God) in Descartes' Physics." *Studies in History and Philosophy of Science* 10 (1979), 113–40.

"The Senses and the Fleshless Eye: The *Meditations* as Cognitive Exercises." In *Essays on Descartes' Meditations.* Edited by Amelie Rorty. Berkeley: University of California Press, 1986.

"The Workings of the Intellect: Mind and Psychology." In *Logic and the Workings of the Mind: The Logic of Ideas and Faculty Psychology in Early Modern Philosophy.* Edited by Patricia Easton. Atascadero, CA: Ridgeview, 1997.

"The Cognitive Faculties." In *The Cambridge History of Seventeenth-Century Philosophy.* Edited by Daniel Garber and Michael Ayers. Cambridge: Cambridge University Press, 1998.

Headley, John M. *Tommaso Campanella and the Transformation of the World.* Princeton: Princeton University Press, 1997.

Henry, John. "Occult Qualities and the Experimental Philosophy: Active Principles in Pre-Newtonian Matter Theory." *History of Science* 24 (1986), 335–81.

"Void Space, Mathematical Realism and Francesco Patrizi da Cherso's Use of Atomistic Arguments." In *Late Medieval and Early Modern Corpuscular Matter Theories.* Edited by Christoph Lüthy, John Murdoch, and William R. Newman. Leiden: Brill, 2001.

Hervey, Helen. "Hobbes and Descartes in the Light of some Unpublished Letters of the Correspondence between Sir Charles Cavendish and Dr. John Pell". *Osiris* 10 (1952), 67–90.

Hirai, Hiro. "Concepts of Seeds and Nature in the Work of Marsilio Ficino." In *Marsilio Ficino: His Theology, His Philosophy, His Legacy.* Edited by Michael J. B. Allen and Valery Rees. Leiden: Brill, 2002.

Holden, Thomas. *The Architecture of Matter: Galileo to Kant.* Oxford: Clarendon Press, 2004.

Huffman, William H. *Robert Fludd and the End of the Renaissance.* London: Routledge, 1988.

Hutcheson, Keith. "What Happened to Occult Qualities in the Scientific Revolution?" *Isis* 73 (1982), 233–53.

James, Susan. "Certain and Less Certain Knowledge." *Proceedings of the Aristotelian Society* 87 (1987), 227–42.

Passion and Action: The Emotions in Seventeenth-Century Philosophy. Oxford: Clarendon Press, 1997.

Jardine, Nicolas. *The Birth of History and Philosophy of Science: Kepler's* A Defence of Tycho against Ursus; *with Essays on Its Provenance and Significance.* Cambridge: Cambridge University Press, 1984.

Jones, Howard. *Pierre Gassendi, 1592–1655: An Intellectual Biography.* Nieuwkoop: B. De Graaf, 1981.

The Epicurean Tradition. London: Routledge, 1989.

"Gassendi and Locke on Ideas." In *Essays on Early Modern Philosophers: Grotius to Gassendi.* Edited by V. C. Chappell. New York: Garland, 1992.

Joy, Lynn Sumida. *Gassendi the Atomist: Advocate of History in an Age of Science.* Cambridge: Cambridge University Press, 1987.

"The Conflict of Mechanisms and Its Empiricist Outcome." *Monist* 71 (1988), 498–514.

"Epicureanism in Renaissance Moral and Natural Philosophy." *Journal of the History of Ideas* 53.4 (1992), 573–83.

Kahn, Didier. "Entre atomisme, alchimie et théologie: la réception des thèses d'Antoine de Villon et Étienne de Clave contre Aristote, Paracelse et les 'cabalistes'." *Annals of Science* 58 (2000): 241–86.

Kargon, Robert. *Atomism in England from Hariot to Newton.* Oxford: Clarendon Press, 1966.

Kessler, Eckhardt. "The Intellective Soul." In *The Cambridge History of Renaissance Philosophy.* Edited by C. B. Schmitt and Quentin Skinner. Cambridge: Cambridge University Press, 1988.

Kirsop, W. "Prolegomènes à une Étude de la Publication et de la Diffusion des Opera Omnia de Gassendi." In *Materia actuosa: antiquité, age classique, Lumières: mélanges en l'honneur d'Olivier Bloch.* Edited by M. Benitez. Paris: H. Chamption, 2000.

Koyré, Alexandre. *Études galiléennes.* Paris: Hermann, 1939.

Newtonian Studies. London: Chapman and Hall, 1965.

Galileo Studies. Translated by John Mepham. New Jersey: Humanities Press, 1978.

From the Closed World to the Infinite Universe. Baltimore: Johns Hopkins University Press, 1994.

Kroll, Richard. "The Question of Locke's Relation to Gassendi." *Journal of the History of Ideas* 45 (1984), 339–60.

The Material Word: Literate Culture in the Restoration and Early Eighteenth Century. Baltimore: Johns Hopkins University Press, 1991.

Leijenhorst, Cees. *The Mechanisation of Aristotelianism: The Late Aristotelian Setting of Thomas Hobbes' Natural Philosophy.* Leiden: Brill, 2002.

Lennon, Thomas. "The Epicurean New Way of Ideas: Gassendi, Locke, and Berkeley." In *Atoms, Pneuma and Tranquility: Epicurean and Stoic Themes in European Thought.* Edited by Margaret J. Osler. Cambridge: Cambridge University Press, 1991.

The Battle of the Gods and the Giants: The Legacy of Descartes and Gassendi, 1655–1715. Princeton: Princeton University Press, 1993.

"Pandora, or, Essence and Reference: Gassendi's Nominalist Objection and Descartes' Realist Reply." In *Descartes and His Contemporaries: Meditations,*

Objections, and Replies. Edited by Roger Ariew and Marjorie Grene. Chicago: University of Chicago Press, 1993.

Lindberg, David. *Theories of Vision from al-Kindi to Kepler.* Chicago: University of Chicago Press, 1976.

LoLordo, Antonia. "The Activity of Matter in Gassendi's Physics." *Oxford Studies in Early Modern Philosophy* 2 (2005), 75–104.

"'Descartes' One Rule of Logic': Gassendi's Critique of Clear and Distinct Perception." *British Journal for the History of Philosophy* 13.1 (2005), 51–72.

"Gassendi on Human Knowledge of the Mind." *Archiv für Geschichte der Philosophie* 87.1 (2005), 1–21.

Long, A. A. *Hellenistic Philosophy: Stoics, Epicureans, Sceptics.* London: Duckworth, 1974.

Lüthy, Christoph. "Thoughts and Circumstances of Sebastien Basso: Analysis, Micro-History, Questions." *Early Science and Medicine* 2.1 (1997), 1–73.

"The Fourfold Democritus on the Stage of Early Modern Science." *Isis* 91.3 (2000), 443–79.

Lüthy, Christoph. "An Aristotelian Watchdog as Avant-Garde Physicist: Julius Caesar Scaliger." *The Monist* 84.4 (2001), 542–61.

MacIntosh, John J. "Robert Boyle on Epicurean Atheism and Atomism." In *Atoms, Pneuma and Tranquility: Epicurean and Stoic Themes in European Thought.* Edited by Margaret J. Osler. Cambridge: Cambridge University Press, 1991.

Mancosu, Paolo. "Aristotelian Logic and Euclidean Mathematics: 17th Century Developments of the Quaestio de certitudine mathematicarum." *Studies in History and Philosophy of Science* 23.3 (1992), 241–65.

Mancosu, Paolo, and Ezio Vailati. "Torricelli's Infinitely Long Solid and Its Philosophical Reception in the 17th Century." *Isis* 82 (1991), 50–70.

Manzo, S. A. "Francis Bacon and Atomism: A Reappraisal." In *Late Medieval and Early Modern Corpuscular Matter Theory.* Edited by Christoph Lüthy, John Murdoch, and William Newman. Leiden: Brill, 2001.

Mayo, Thomas F. *Epicurus in England (1650–1725).* Dallas: The Southwest Press, 1934.

Mazaurac, Simone. *Gassendi, Pascal et la querelle du vide.* Paris: Presses Universitaires de France, 1998.

Meinel, Christoph. "Early Seventeenth-Century Atomism: Theory, Epistemology, and Insufficiency of Experiment." *Isis* 79 (1988), 68–103.

"Empirical Support for the Corpuscular Philosophy in the Seventeenth Century." In *Theory and Experience: Recent Insights and New Perspectives on their Relation.* Edited by Diderik Batens and Jean Paul van Bendegem. Dordrecht: Reidel, 1988.

Mendelsohn, Everett. *Heat and Life: The Development of the Theory of Animal Heat.* Cambridge, MA: Harvard University Press, 1964.

Menn, Stephen. *Descartes and Augustine.* Cambridge: Cambridge University Press, 1998.

Michael, Emily. "Two Early Modern Concepts of Mind: Reflecting Substance vs. Thinking Substance." *Journal of the History of Philosophy* 27 (1989), 29–48.

"The Theory of Ideas in Gassendi and Locke." *Journal of the History of Ideas* 51 (1990), 379–99.

"Daniel Sennert on Matter and Form: At the Juncture of the Old and the New." *Early Science and Medicine* 2.3 (1997), 272–99.

"Sennert's Sea Change: Atoms and Causes." In *Late Medieval and Early Modern Corpuscular Matter Theories*. Edited by Christoph Lüthy, John Murdoch, and William Newman. Leiden: Brill, 2001.

Michael, Emily, and Fred Michael. "Gassendi on Sensation and Reflection: A Non-Cartesian Dualism." *History of European Ideas* 9 (1988), 583–95.

"Corporeal Ideas in Seventeenth-Century Psychology." *Journal of the History of Ideas* 50 (1989), 31–48.

"A Note on Gassendi in England." *Notes and Queries* 37.3 (1990), 297–9.

"Gassendi's Modified Epicureanism and British Moral Philosophy." *History of European Ideas* 21.6 (1995), 743–61.

Michael, Fred. "Why Logic Became Epistemology: Gassendi, Port Royal and the Reformation in Logic." In *Logic and the Workings of the Mind: The Logic of Ideas and Faculty Psychology in Early Modern Philosophy*. Edited by Patricia Easton. Atascadero, CA: Ridgeview, 1997.

Miller, Peter N. *Peiresc's Europe: Learning and Virtue in the Seventeenth Century*. New Haven: Yale University Press, 2000.

Möll, Konrad. *Der junge Leibniz*. 3 volumes. Stuttgart-Bad Cannstatt: Frommann-Holzboog, 1978–1996.

Monnoyeur, Françoise. "Matter: Descartes versus Gassendi." Presentation to the University of Virginia philosophy department, October 2001. To be published as part of her *Queen Christina and the Sciences of Her Time*. Paris: Herman Cohen, forthcoming.

Murdoch, John E. "Infinity and Continuity." In *The Cambridge History of Later Medieval Philosophy*. Edited by Norman Kretzmann, Anthony Kenny, Jan Pinborg, and Eleanore Stump. Cambridge: Cambridge University Press, 1982.

Murdoch, John E. "The Medieval and Renaissance Tradition of *Minima Naturalia*." In *Late Medieval and Early Modern Corpuscular Matter Theory*. Edited by Christoph Lüthy, John Murdoch, and William Newman. Leiden: Brill, 2001.

Murr, Sylvia. "Préliminaires à la Physique *Syntagma Philosophicum*." *Dix-septième siècle* 45.2 (1993), 353–485.

Nadler, Steven. "Doctrines of Explanation in Late Scholasticism and the Mechanical Philosophy." In *The Cambridge History of Seventeenth-Century Philosophy*. Edited by Daniel Garber and Michael Ayers. Cambridge: Cambridge University Press, 1998.

Newman, William R. "The Alchemical Sources of Robert Boyle's Corpuscular Philosophy." *Annals of Science* 53 (1996), 567–85.

"Experimental Corpuscular Theory in Aristotelian Alchemy: From Geber to Sennert." In *Late Medieval and Early Modern Corpuscular Matter Theory*. Edited by Christoph Lüthy, John Murdoch, and William Newman. Leiden: Brill, 2001.

Newman, William R., and Larry Principe. "Alchemy vs. Chemistry: The Etymological Origins of a Historiographical Mistake." *Early Science and Medicine* 3.1 (1998), 32–65.

Newman, William R., and Lawrence Principe. *Alchemy Tried in the Fire*. Chicago: University of Chicago Press, 2002.

Normore, Calvin. "Meaning and Objective Being: Descartes and His Sources." In *Essays on Descartes' Meditations.* Edited by Amelie Rorty. Berkeley: University of California Press, 1986.

Norton, David Fate. "The Myth of 'British Empiricism'." *American Philosophical Quarterly* 1 (1981), 331–44.

Nuchelmans, Gabriel. *Late-Scholastic and Humanistic Theories of the Proposition.* Amsterdam: North-Holland, 1980.

Judgment and Proposition from Descartes to Kant. Amsterdam: North-Holland, 1983.

Osler, Margaret J. "Providence and Divine Will: The Theological Background to Gassendi's Views on Scientific Knowledge." *Journal of the History of Ideas* 44 (1983), 549–60.

"Baptizing Epicurean Atomism: Pierre Gassendi on the Immortality of the Soul." In *Religion, Science, and Worldview: Essays in Honor of Richard S. Westfall.* Edited by Margaret J. Osler and Paul Lawrence Farber. Cambridge: Cambridge University Press, 1985.

Osler, Margaret J. "Fortune, Fate, and Divination: Gassendi's Voluntarist Theology and the Baptism of Epicureanism." In *Atoms, Pneuma and Tranquility: Epicurean and Stoic Themes in European Thought.* Edited by Margaret J. Osler. Cambridge: Cambridge University Press, 1991.

"The Intellectual Sources of Robert Boyle's Philosophy of Nature: Gassendi's Voluntarism and Boyle's Physico-Theological Project." In *Philosophy, Science and Religion in England, 1640–1700.* Edited by Richard Kroll. Cambridge: Cambridge University Press, 1992.

"Ancients, Moderns and the History of Philosophy: Gassendi's Epicurean Project." In *The Rise of Modern Philosophy: The Tension between the New and Traditional Philosophies from Machiavelli to Leibniz.* Edited by Tom Sorell. Oxford: Clarendon Press, 1993.

Divine Will and the Mechanical Philosophy. Cambridge: Cambridge University Press, 1994.

"Divine Will and Mathematical Truth: Gassendi and Descartes on the Status of Eternal Truths." In *Descartes and His Contemporaries: Meditations, Objections, and Replies.* Edited by Roger Ariew and Marjorie Grene. Chicago: University of Chicago Press, 1995.

"How Mechanical Was the Mechanical Philosophy? Non-Epicurean Aspects of Gassendi's Philosophy of Nature." In *Late Medieval and Early Modern Corpuscular Matter Theory.* Edited by Christoph Lüthy, John Murdoch, and William Newman. Leiden: Brill, 2001.

"Whose Ends? Teleology in Early Modern Philosophy." *Osiris* 16 (2001), 151–68.

"The History of *Philosophy* and the *History* of Philosophy: A Plea for Textual History in Context." *Journal of the History of Philosophy* 40.4 (2002), 529–33.

Pagel, Walter. *Joan Baptista van Helmont: Reformer of Science and Medicine.* Cambridge: Cambridge University Press, 1982.

Palmerino, Carla Rita. "Pierre Gassendi's *De Philosophia Epicuri Universi* Rediscovered." *Nuntius* 14 (1998): 131–62.

"Infinite Degrees of Speed: Marin Mersenne and the Debate over Galileo's Law of Free Fall." *Early Science and Medicine* 4 (1999): 269–328.

"Galileo's and Gassendi's Solution to the *Rota Aristotelis* Paradox: A Bridge between Matter and Motion Theories." In *Late Medieval and Early Modern Corpuscular Matter Theory.* Edited by Christoph Lüthy, John Murdoch, and William Newman. Leiden: Brill, 2001.

"Two Jesuit Responses to Galileo's Science of Motion: Honoré Fabri and Pierre Le Cazré." In *The New Science and Jesuit Science: Seventeenth Century Perspectives.* Edited by Mordechai Feingold. Dordrecht: Kluwer, 2003.

"Gassendi's Reinterpretation of the Galilean Theory of Tides." *Perspectives on Science* 12.2 (2004): 212–37.

Palmerino, Carla Rita, and J. M. M. H. Thijssen. *The Reception of the Galilean Science of Motion in Seventeenth-Century Europe.* Dordrecht: Kluwer, 2004.

Panchieri, Lillian Unger. "The Magnet, the Oyster, and the Ape, or Pierre Gassendi and the Principle of Plenitude." *Modern Schoolman* 53 (1976), 141–50.

Panchieri, Lillian Unger. "Pierre Gassendi: A Forgotten but Important Man in the History of Physics." *American Journal of Physics* 46 (1978), 435–63.

Pasnau, Robert. *Theories of Cognition in the Later Middle Ages.* Cambridge: Cambridge University Press, 1997.

Pav, Peter. "Gassendi's Statement of the Principle of Inertia." *Isis* 57 (1966), 23–34.

Pintard, René. *Le libertinage érudit dans la première moitié du XVII siècle.* 2 volumes. Paris: Boivin, 1943.

La Mothe le Vayer–Gassend–Guy Patin: Études de bibliographie et de critique suivies de textes inédits de Guy Patin. Paris: Boivin, 1943.

Popkin, Richard. *The History of Scepticism: From Savonarola to Bayle.* Oxford: Oxford University Press, 2003.

Principe, Lawrence. *The Aspiring Adept: Robert Boyle and His Alchemical Quest.* Princeton: Princeton University Press, 1998.

Pyle, Andrew. *Atomism and Its Critics: Problem Areas Associated with the Development of the Atomic Theory of Motion from Democritus to Newton.* Bristol: Thoemmes, 1995.

Rochot, Bernard. *Les travaux de Gassendi sur Epicure et sur l'atomisme, 1619–1658.* Paris: J. Vrin, 1944.

"Gassendi et la 'logique' de Descartes." *Revue philosophique de la France et de l'étranger* 141 (1951), 288–98.

"Chronologie de la Vie et des Ouvrages de Pierre Gassendi." In *Pierre Gassendi 1592–1655: Sa Vie et son Oeuvre.* Edited by Bernard Rochot. Paris: Editions Albin Michel, 1955.

"Vie et Caractère." In *Pierre Gassendi 1592–1655: Sa Vie et son Oeuvre.* Edited by Bernard Rochot. Paris: Editions Albin Michel, 1955.

"Comment Gassendi interprétait l'expérience du Puy de Dôme." *Revue d'histoire des sciences* 16 (1963), 53–76.

Rodis-Lewis, Geneviève. *Descartes: His Life and Thought.* Translated by Jane Marie Todd. Ithaca, NY: Cornell University Press, 1998.

Rogers, G. A. J. "Gassendi and the Birth of Modern Philosophy." *Studies in History and Philosophy of Science Part A* 26.4 (1995), 681–7.

Sarasohn, Lisa T. "The Ethical and Political Philosophy of Pierre Gassendi." *Journal of the History of Philosophy* 20 (1982), 239–60.

"Motion and Morality: Pierre Gassendi, Thomas Hobbes, and the Mechanical World-View." *Journal of the History of Ideas* 46 (1985): 363–80.

"French Reaction to the Condemnation of Galileo, 1632–1642." *Catholic Historical Review* 74 (1988): 34–54.

"Nicolas-Claude Fabri de Peiresc and the Patronage of the New Science." *Isis* 84 (1993), 70–90.

Gassendi's Ethics: Freedom in a Mechanistic Universe. Ithaca, NY: Cornell University Press, 1996.

Schmitt, Charles B. "The Rediscovery of Ancient Skepticism in Modern Times." In *The Sceptical Tradition.* Edited by Myles Burnyeat. Berkeley: University of California Press, 1983.

The Aristotelian Tradition and Renaissance Universities. London: Variorum Reprints, 1984.

Schneewind, J. B. *The Invention of Autonomy: A History of Modern Moral Philosophy.* Cambridge: Cambridge University Press, 1998.

Secada, Jorge. *Cartesian Metaphysics: The Late Scholastic Origins of Modern Philosophy.* Cambridge: Cambridge University Press, 2000.

Sedley, David. *Lucretius and the Transformation of Greek Wisdom.* Cambridge: Cambridge University Press, 1998.

Shackelford, Joel. "Seeds with a Mechanical Purpose." In *Reading the Book of Nature.* Edited by A. G. Debus and M. T. Walton. Kirksville, MO: Sixteenth Century Journal Publishers, 1998.

Shapin, Steven, and Simon Schaffer. *Leviathan and the Air-Pump: Hobbes, Boyle, and the Experimental Life.* Princeton: Princeton University Press, 1985.

Simmons, Alison. "Explaining Sense Perception: A Scholastic Challenge." *Philosophical Studies* 73.2–3 (1994), 257–75.

Sorabji, Richard. *Time, Creation, and the Continuum: Theories in Antiquity and the Early Middle Ages.* Ithaca, NY: Cornell University Press, 1983.

Matter, Space, and Motion: Theories in Antiquity and Their Sequel. Ithaca, NY: Cornell University Press, 1988.

Sorrell, Tom. "Seventeenth-Century Materialism: Gassendi and Hobbes." In *The Renaissance and 17th Century Rationalism.* Edited by G. H. R. Parkinson. New York: Routledge, 1993.

Sortais, Gaston. *La philosophie moderne depuis Bacon jusqu'à Leibniz.* 2 volumes. Paris: Paul Lethielleux, 1920–2.

Spink, J. S. *French Free-thought from Gassendi to Voltaire.* New York: Greenwood Press, 1969.

Thorndike, L. *A History of Magic and Experimental Science.* New York: Columbia University Press, 1947.

Vlastos, Gregory. "Minimal Parts in Epicurean Atomism." *Isis* 56 (1965), 121–47.

Walker, Ralph. "Gassendi and Skepticism." In *The Sceptical Tradition.* Edited by Myles Burnyeat. Berkeley: University of California Press, 1983.

Wallace, William. *Prelude to Galileo: Essays on Medieval and Sixteenth-Century Sources of Galileo's Thought.* Dordrecht: Reidel, 1981.

Westfall, Richard S. *Force in Newton's Physics: The Science of Dynamics in the Seventeenth Century*. New York: American Elsevier, 1971.

Wilson, Catherine. *The Invisible World: Early Modern Philosophy and the Invention of the Microscope*. Princeton: Princeton University Press, 1995.

"Theological Foundations for Modern Science?" *Dialogue* 36 (1997), 597–606.

Yolton, John. *Perceptual Acquaintance from Descartes to Reid*. Minneapolis: University of Minnesota Press, 1984.

Perception and Reality: A History from Descartes to Kant. Ithaca, NY: Cornell University Press, 1996.

Index